M000306331

Tourism and Souvenirs

TOURISM AND CULTURAL CHANGE
Series Editors: Professor Mike Robinson, *The Ironbridge International Institute for Cultural Heritage, University of Birmingham, UK* and Dr Alison Phipps, *University of Glasgow, Scotland, UK.*

Understanding tourism's relationships with culture(s) and vice versa, is of ever-increasing significance in a globalising world. This series will critically examine the dynamic inter-relationships between tourism and culture(s). Theoretical explorations, research-informed analyses and detailed historical reviews from a variety of disciplinary perspectives are invited to consider such relationships.

Full details of all the books in this series and of all our other publications can be found on http://www.channelviewpublications.com, or by writing to Channel View Publications, St Nicholas House, 31–34 High Street, Bristol BS1 2AW, UK.

Tourism and Souvenirs

Glocal Perspectives from the Margins

Edited by
Jenny Cave, Lee Jolliffe and Tom Baum

CHANNEL VIEW PUBLICATIONS
Bristol • Buffalo • Toronto

Library of Congress Cataloging in Publication Data
A catalog record for this book is available from the Library of Congress.
Tourism and Souvenirs: Glocal Perspectives from the Margins/Edited by Jenny Cave, Lee Jolliffe and Tom Baum.
Tourism and Cultural Change: 33
Includes bibliographical references and index.
1. Tourism—Social aspects. 2. Souvenirs (Keepsakes)—Social aspects. I. Cave, Jenny.
G155.A1T58945 2013
306.4'819–dc23 2013011778

British Library Cataloguing in Publication Data
A catalogue entry for this book is available from the British Library.

ISBN-13: 978-1-84541-406-1 (hbk)
ISBN-13: 978-1-84541-405-4 (pbk)

Channel View Publications
UK: St Nicholas House, 31–34 High Street, Bristol BS1 2AW, UK.
USA: UTP, 2250 Military Road, Tonawanda, NY 14150, USA.
Canada: UTP, 5201 Dufferin Street, North York, Ontario M3H 5T8, Canada.

The policy of Multilingual Matters/Channel View Publications is to use papers that are natural, renewable and recyclable products, made from wood grown in sustainable forests. In the manufacturing process of our books, and to further support our policy, preference is given to printers that have FSC and PEFC Chain of Custody certification. The FSC and/or PEFC logos will appear on those books where full certification has been granted to the printer concerned.

Typeset by Techset Composition India (P) Ltd., Bangalore and Chennai, India.
Printed and bound in Great Britain by Short Run Press Ltd.

Contents

Contributors vii

1 Theorising Tourism and Souvenirs, Glocal Perspectives
  on the Margins 1
 Jenny Cave, Tom Baum and Lee Jolliffe

Part 1: Theorising Experience and Behaviour

2 With the Passing of Time: The Changing Meaning of Souvenirs 29
 Noga Collins-Kreiner and Yael Zins

3 Souvenirs and Self-identity 40
 Hugh Wilkins

4 Souveniring Occupational Artefacts: The Chef's Uniform 49
 Richard N.S. Robinson

Part 2: Theorising Place and Identity

5 Souvenirs of the American Southwest: Objective or Constructive
  Authenticity? 63
 Kristen K. Swanson

6 'Souvenirs' at the Margin? Place, Commodities, Transformations
  and the Symbolic in Buddha Sculptures from Luang
  Prabang, Laos 82
 Russell Staiff and Robyn Bushell

7 Souvenirs as Transactions in Place and Identity: Perspectives from
  Aotearoa New Zealand 98
 Jenny Cave and Dorina Buda

Part 3: Glocal Case Studies in Sustainable Tourism

8 Green Tourism Souvenirs in Rural Japan: Challenges and
    Opportunities   119
   Atsuko Hashimoto and David J. Telfer

9 Understanding Tourist Shopping Village Experiences on
    the Margins   132
   Laurie Murphy, Gianna Moscardo and Pierre Benckendorff

10 Souvenir Development in Peripheral Areas: Local Constraints in
    a Global Market   147
   R. Geoffrey Lacher and Susan L. Slocum

11 Souvenir Production and Attraction: Vietnam's Traditional
    Handicraft Villages   161
   Huong T. Bui and Lee Jolliffe

12 World Heritage-themed Souvenirs for Asian Tourists in Macau   176
   Hilary du Cros

13 Lessons in Tourism and Souvenirs from the Margins:
    Glocal Perspectives   189
   Lee Jolliffe, Jenny Cave and Tom Baum

Index   200

Contributors

Tom Baum is Programme Director, Hong Kong University SPACE programmes in Tourism and Hospitality Management at the University of Strathclyde. He is also a Board Member of the Economic Research Institute for Northern Ireland. His research agenda includes: people and work in low skills service work, with a particular focus on the international hospitality and tourism sector as well as human resource development and skills planning and formation, education and training, at a macro (national) and company level. He also examines migration and wider mobilities in the low skills service work sector.

Dr Pierre Benckendorff is a Senior Lecturer and social scientist in the School of Tourism, The University of Queensland, Australia. He has more than 10 years of experience in education and research in the tourism field in Australia and internationally. His research interests include consumer behaviour, the impact of new technologies on tourism, tourism education and tourism scholarship and epistemology. Contact: p.benckendorff@uq.edu.au.

Dr Dorina Maria Buda is a Senior Lecturer in Tourism at Saxion Hospitality Business School in The Netherlands. Dorina examines interconnections between tourism and socio-political conflicts as she concentrates on tourist subjectivities, emotions, feelings, affect and senses. In her critical and qualitative methodological approach to tourism studies, Dorina employs a wide range of methods from individual and group interviews to written- and photo-diaries. She completed a doctoral programme in Tourism and Geography at the University of Waikato in Aotearoa New Zealand.

Huong T. Bui has a PhD in Tourism Management from Griffith University, Australia and is currently an Assistant Professor in Tourism at Ritsumeikan Asia Pacific University, Japan. Her research interest is youth tourism in

Asia and tourism development in Southeast Asia. She is a former consultant for tourism development projects in Vietnam.

Robyn Bushell is Associate Professor in the Institute for Culture and Society, University of Western Sydney. Her work focuses on the values underpinning everyday life, quality of life, sustainable development and heritage management. She works closely with a range of national and international heritage conservation organisations in the formulation policies and planning frameworks, in particular community development strategies involving tourism. Her current research interests examine the entangled relationships between the local/global and between conservation and development in heritage places. She has recently co-edited a volume on Heritage and Tourism for Routledge.

Jenny Cave is a Senior Lecturer in Tourism and Hospitality Management at the University of Waikato, New Zealand. Her background in anthropology, museology, tourism and cultural attraction management shapes a research agenda in the linkages between tourism, migration and poverty reduction in rural and island peripheries, as well as cultural/heritage enterprise and collaborative methodologies. She has published on backpacker tourism and investment/yield macro-trends in the travel and leisure sector. Jenny has been Visiting Scholar at the Cairns Institute, James Cook University, Australia, the University of the West Indies, Barbados, and the University of the South Pacific (Fiji and Tonga). She serves on the editorial boards of the *International Journal of Event and Festival Management* and the *Journal of Tourism* and has guest-edited two volumes on island tourism (marketing heritage and destinations) for the *International Journal of Culture, Tourism and Hospitality Research*.

Noga Collins-Kreiner is a Senior Lecturer (PhD), in the Department of Geography and Environmental Studies at the University of Haifa, Israel (http://geo.haifa.ac.il/~nogack/) and a member of the Center for Tourism, Pilgrimage & Recreation Research at the University of Haifa, Israel (http://ctprr.haifa.ac.il). Her main research interests are: tourism, pilgrimage and tourism development and management. She is also a resource editor of *Annals of Tourism Research* and published many papers on the topic of tourism and the *Christian Tourism to the Holy Land: Pilgrimage During Security Crisis* (2006, Ashgate). One of her latest papers is: Collins-Kreiner N. (2010) Researching pilgrimage: Continuity and transformations. *Annals of Tourism Research* 37 (2), 440–456.

Hilary du Cros is currently engaged as an Associate Professor at the Hong Kong Institute of Education. She has taught and has worked in the Asia

Pacific region over the last 28 years (including projects for the United Nations World Tourism Organization or UNESCO). She has an interdisciplinary perspective on heritage and arts management, marketing and sustainable tourism development. Dr du Cros has published over 118 works that include a book with Yok-shiu F. Lee, *Cultural Heritage Management in China* (2007) and the popular textbook with Bob McKercher, *Cultural Tourism: The Partnership between Tourism and Cultural Heritage Management* (2002; 2nd edn, 2014).

Atsuko Hashimoto is Associate Professor in the Department of Tourism and Environment at Brock University. She is also an Associate Member of the Environmental Sustainability Research Centre at Brock. Her research focuses on sociocultural, intercultural and human aspects of tourism development and green tourism in Japan.

Lee Jolliffe is a Professor at the University of New Brunswick in Canada. With an academic background in museology she has research interests in the intersection of culture and tourism, especially in urban and rural settings, as well as museums and arts events in North Atlantic Islands, South East Asia and the Caribbean. Internationally she has been a Visiting Professor at Hanoi University, Vietnam and the Almond Chair in Tourism and Hospitality Management at the University of West Indies, Barbados. Her edited volumes include *Tea and Tourism: Tourists, Traditions and Transformations* (Channel View Publications, 2007) and *Coffee Culture and Destinations* (Channel View Publications, 2010). Lee serves on the editorial boards of *Annals of Tourism Research*, the *International Journal of Contemporary Hospitality Management and Tourism* and the *International Journal of Tourism Policy*.

R. Geoffrey Lacher is an Assistant Professor in the School of Community Resources and Development at Arizona State University. He received his PhD from Clemson University's Department of Parks, Recreation and Tourism Management. His research focuses on the economics on tourism with an emphasis on sustainable tourism and tourism in the developing world. He has previous research experience in Thailand, the Dominican Republic and Tanzania, as well as the United States.

Professor Gianna Moscardo has qualifications in applied psychology and sociology and joined the School of Business at James Cook University in 2002. Prior to joining JCU Gianna was the Tourism Research project leader for the CRC Reef Research for eight years. Her research interests include understanding how consumers, especially, tourists make decisions and evaluate their experiences, and how communities and organisations perceive, plan for, and manage tourism development opportunities. She has published extensively on tourism and related areas with more than 170 refereed papers or book chapters. Contact: gianna.moscardo@jcu.edu.au.

Associate Professor Laurie Murphy lectures in the School of Business at James Cook University in the area of tourism, events and sports management. Laurie's research interests focus on improving tourism's contribution to regional communities with an emphasis on tourism marketing, including a focus on the backpacker market, destination image and choice, and more recently destination branding and tourist shopping villages. Laurie is on the editorial board of both the Journal of Travel Research and the Journal of Travel and Tourism Marketing and serves on the Tourism Development and Marketing Strategic Advisory Committee for Townsville Enterprise Ltd. Contact: Laurie.Murphy@jcu.edu.au.

Richard Robinson is Senior Lecturer in Hospitality Management at the University of Queensland, School of Tourism. His research focuses on tourism and hospitality industry workforce issues, food tourism and the scholarship of teaching and learning. He has coordinated and worked on research teams for funded national and international projects relating to skills development, occupational cultures, mobility and workforce policy issues, investigating the characteristics and consumer behaviour of highly involved 'foodies' and designing and evaluating education programmes in tourism and hospitality. His work in these areas has been published in leading academic journals, edited books, international conference proceedings and practitioner periodicals.

Susan L. Slocum specialises in sustainable economic development through tourism and policy implementation at the regional and national level. Working with communities to enhance backward linkages between tourism and traditional industries, Dr Slocum has worked with rural communities across the UK to develop food tourism initiatives and with indigenous populations in emerging tourism destinations in Tanzania. In particular, she is interested in balancing policy development and integration to provide a more bottom-up form of planning within tourism destinations and has approached sustainable tourism from a contemporary view that includes the addition of institutional reform and social justice. Dr Slocum is currently employed by the Institute for Tourism Research at the University of Bedfordshire in the UK.

Russell Staiff holds a doctorate in art history from the University of Melbourne and is a member of the Institute of Culture and Society within the University of Western Sydney where he researches the interface between cultural heritage, tourism and communities with a special focus on Southeast Asia. He has recently co-edited a volume on heritage and tourism for Routledge and a monograph on re-imagining heritage interpretation is forthcoming from Ashgate.

Kristen K. Swanson is Professor at Northern Arizona University, Flagstaff, Arizona, in the School of Communication. Her research interests focus on souvenirs, tourism retailing and shopping behaviour; tourism in the American Southwest; and fashion promotion and events. Her most recent publication on souvenirs has appeared in the scholarly journal *Tourism Management*.

David J. Telfer is Associate Professor in the Department of Tourism and Environment at Brock University. He is also an Associate Member of the Environmental Sustainability Research Centre at Brock. His research focuses on tourism and development theory, the linkages between tourism and agriculture and green tourism in Japan.

Hugh Wilkins is a Professor at Edith Cowan University in Western Australia and is Head of the School of Marketing, Tourism & Leisure. His teaching and research interests are in consumer behaviour, marketing and strategic management in the hotel, hospitality and tourism industries. Prior to joining Edith Cowan University he worked at Griffith University in Queensland where he had varying roles including Director, Offshore Operations and Head, School of Tourism and Hotel Management. He moved to Australia in 1996 and before that taught in the UK at Oxford Brookes and Staffordshire Universities.

Yael Zins has graduated with an MA degree in planning and developing tourism resources in the Department of Geography and Environmental Studies at the University of Haifa. The topic of her MA thesis is souvenir acquisition patterns and meanings of souvenirs. Today she works in the tourism industry.

1 Theorising Tourism and Souvenirs, Glocal Perspectives on the Margins

Jenny Cave, Tom Baum and Lee Jolliffe

Around the globe, the production of souvenirs (supply) and participation in their acquisition (consumption) play central roles in actively sustaining tourism economies, community relationships and cultural structures, traditions and heritage. This book situates souvenirs as tangible and intangible expressions and triggers of tourism experience that are 'glocally' developed on the margins, at tourism peripheries.

Within the Channel View series of volumes concerning tourism's role in globalisation and cultural change, this book explores the processes and particularities of glocal (combining global and local) construction of souvenirs of place, people and experiences. It examines souvenirs as glocal agents in resisting, responding and interpreting global influences at local levels – preserving and sustaining craft traditions, cultural structures, community relationships and economies located on geographic, cultural, political, societal and economic margins of the globe.

Situated within a theorisation of glocalisation and its impacts, separate but interconnected issues are addressed in this volume. These include: the role of people, traditions and relationships; the role of place and symbolism, authenticity, production and material culture; the management of experiences at and/or near destinations; product marketing; human resource management issues of labour, intermediaries, yield and sales; and community development, cultural entrepreneurship and governmental initiatives. The book models the global spread of the three editors, who bring southern and northern hemispheric perspectives to the volume, as well as personal research interests in areas that include island and peripheral area tourism, mobility, human resources, heritage, entrepreneurship and material culture that span several continents and islands (Africa, Asia, Australasia, the Caribbean, the Pacific Islands, North America, the United Kingdom, Scandinavia and elsewhere in Europe).

Publication of this volume is timely since the only other dedicated volume on this subject, *Souvenirs: The Material Culture Of Tourism* by M. Hitchcock and K. Teague (2000) was published more than a decade ago. This multidisciplinary volume aims to complement the primarily anthropological work offered by that volume. First we review the literature on souvenirs as a context for extending the research agenda on souvenirs.

Souvenirs – A Brief Review

The souvenir research field has been surveyed by Swanson and Timothy (2012) who include an analysis of the demand and supply of production, consumption and distribution, suggesting a taxonomy of tangible and intangible souvenirs as symbolic reminders and commodities messengers of meaning, souvenirs, tradable commodities and commodification. A small number of volumes deal with related topics, but the focus is more in terms of patterns of consumption through retail shopping (Timothy, 2005) and tourist shopping villages (Murphy *et al.*, 2011), while Lippard (1999) writes on the related subject of domestic tourism and its impacts on local people, art and place, primarily in the United States. Jules-Rosette (1984) analyses tourist art in Africa and Graburn (1976) complied an important global survey of ethnic and tourists arts in the 'Fourth world', which focuses on aboriginal or native peoples whose lands fall within the national boundaries and techno-bureaucratic administrations of First, Second and Third World countries. Our volume includes views of souvenirs from politicised peripheries and communities distanced by geographic and social mechanisms.

A handful of research volume chapters have touched upon, but not been wholly devoted to, the topic of souvenirs and their role in tourism (Deitch, 1989; Jolliffe, 2007). Other authors investigate the impacts of tourism on arts and crafts, dynamics of the marketplace and the proactive role of arts and crafts in sustaining traditions, which can then lead to tourism product (Graburn, 1976; Grünewald, 2006; Lury, 1997; Miettinen, 2006; Parris, 1984; Perkins & Morphy, 2006; Rinzin, 2006; Sharpley, 1994).

Several journal articles address the role of souvenirs in society and the formation of identity (Cohen, 1985; Cohen, 1988; Gordon, 1986; Haggard & Williams, 1992; Shamir, 1992), authenticity (Littrell & Anderson, 1993;

Reisinger & Steiner, 2006) and cultural or self-expression (Asplet & Cooper, 2000; Kim & Littrell, 2001), as well as linkages to different forms of tourism (Littrell *et al.*, 1994). Others examine the role of souvenirs in tourism in peripheral areas (Adams, 1997; Cohen, 1983; Forsyth, 1995), and yet others, their importance to community-initiated development (Fuller & Cummings, 2005; Healy, 1994; Hume, 2009; Kleymeyer, 1994; Tosun, 2000), as well as to island contexts (Holder, 1989; Townsend & Cave, 2004). Some look at the multiplicity of forms (Cohen, 1993b, 2001) and their role in the enterprise development (Cave *et al.*, 2007; Timothy & Wall, 1997) and the purchase experience (Anderson & Littrell, 1995; Fairhurst *et al.*, 2007; Kim & Littrell, 1999; Meethan *et al.*, 2006; Turner & Reisinger, 2001; Yu & Littrell, 2005). Wilkins (2011) investigates the influence of gender purchasing behaviour, as well as the meanings of souvenir purchases to the individual in terms of reasons for purchase and planned uses of the item. Swanson and Timothy (2012) have extended the definition of souvenir to include items not manu-factured as tourist mementos, but which nonetheless function as memory holders. They also examine the function of intermediaries and global manufacturers.

A sample of the wider literature of arts and crafts demonstrates how souvenirs, as highly diversified tourist product, have evolved over time from simple crafts (Cohen, 1993b), are influenced in their design by tourist demand (Boynton, 1986) and are significant to local economies. Governments play pivotal roles in stimulating arts and craft production (Hamel, 2001; Healy, 1994), as do external newcomers (Cohen, 1993a). Migrants hybridise art and craft (Hollinshead *et al.*, 2009), but hosts can actively subvert globalisation by creating reactive identities and cultural product (Wherry, 2006) that serve to protect treasures, traditions and reli-gion (Vorlaufer, 1999). Visitor centres can promote and conserve regional identity (Novotny *et al.*, 2008) by being a locus of production to encourage retention of traditional lifestyles and foods (Kumar & Janz, 2010). Perceptions of what is 'traditional' affect the negotiations between artisans and retail intermediaries, as well as goods chosen for the tourist consumer (Moreno & Littrell, 2001).

As editors, we are delighted that this book includes contributions by several of the authors who have already contributed to this emerging subject of souvenirs and tourism, including Cave (Cave, Jolliffe & De Coteau, 2012), du Cros (McKercher & du Cros, 2002), Hashimoto and Telfer (2008), Jolliffe (2007), Swanson and Timothy (2012) and Wilkins (2011). These authors have used the opportunity provided by this volume to extend their thinking to encompass new data and wider horizons. Next we will address the theo-retical underpinnings of this volume, deconstructing the global–local relationship as a research framework for the volume. How does the souvenir literature fit into the theoretical constructs of glocalisation and peripheralities?

Glocal

The global–local relationship refers to a nexus that ties together economic factors and socio-cultural responses to avoid oversimplifying complex social, cultural and economic processes of interactions of objects with place and movements of people into simple dichotomies. The neologism 'glocalisation' was coined simultaneously by sociologist R. Robertson and geographer E. Swyngedouw 'to capture the interdependent relation between local–global by indicating that all socio-spatial processes may be viewed as simultaneously global and local' (Haldrup, 2009: 250). Globalisation emphasises universality of worldwide cultural or corporate processes, while glocalisation accentuates particularisation of product, service or theme (Matusitz, 2010), or global heterogeneity, where the interpenetration of the global and the local result in unique outcomes in different geographical areas (Ritzer, 2003). The interactivity and/or separateness of these concepts can be read as different perspectives and processes that are contingent; that is, each derives meaning from the other, within connected networks that are always in motion and never completed (Gibson-Graham, 2002).

In terms of perspective, the global may be local if it refers to processes that impact on only some parts of the globe. However, the local may also be global as concretely specific locations embedded in specialised networks of social relations that have global reach. Further, places contain processes that can be globalised and replicated in neighbourhoods across the world (Gibson-Graham, 2002). In this book, we see souvenirs as part of glocal tourism transactions. On the global–local continuum, souvenirs of place and identity thus refer to both the universality and contextuality of tourism transactions.

There is a fear that what appears to be an exchange of ideas or phenomena can be unequal, with local lives and livelihoods transformed, politicised and overtaken (Dirlik, 1999). Also it is perceived that the forces of commodification conveyed by globalisation will produce increasing levels of cultural homogenisation and social standardisation in media icons, social styles and consumption values. And these forces may, in turn, erase cultural differences both within and across societies. But to the contrary, there is evidence that societies appropriate technologies, download media idiosyncratically and that local entrepreneurs and customers subvert standardisation to produce brand differentiation (Appadurai, 2001). An example is the subversion of the uniformity of products in each cultural setting of East Asia of the global chain of McDonald's restaurants where the inherent capitalistic sensitivity to local markets meant that the planners were drawn into the local mosaic of social patterns and cultural orientations, which inevitably produced culturally distinct versions, but not in any predictable way (Watson, 1997).

The globalised heterogeneity versus homogeneity debate cannot be seen as a simple dichotomy of global producer equals uniformity; or local consumer equals difference. Increasingly we see complexities played out in global commodities that are locally interpreted by producers as well as consumers, and that cultural differentiation outpaces homogenisation (Appadurai, 2001). Globalising agents are knowledge-carrying individuals (Lowe *et al.*, 2012), cross-cultural trades, religions, multinational companies and transnational networks of global mobility (Pieterse, 1994). Yet technologies, corporate governance and macro-economic policies are also globalising forces that have a complex, interactive and 'multi-nested' relationship with the national, regional and local economic development of tourism (Milne & Ateljevic, 2001).

Impacts of glocalisation

Globalised contemporary life has been observed as having the consequences of homogenisation of different cultures that brings out the commonality of all cultures owing to the heightened contact between them, but it also generates a renewed emphasis on ethnicity and community control, with the consequence of tourism as a community response (Ezarik, 2003) and localised concerns with cultural identity, historical memory and collective belonging (Doorne *et al.*, 2003). Glocalisation thus is a reflexive and resistant way that people at local levels interpret global processes or phenomena to suit their specific cultural contexts; a relational interplay driven by international lifestyle migration that inevitably impacts on social practices and organisation (Torkington, 2012).

Giulianotti and Robertson (2007) identify four types of glocalisation that relate to cultural institutions, practices and meanings. First, relativisation, in which prior cultural practices are retained within a new environment, differentiated from the new culture, and second, accommodation, which is the pragmatic absorption of practices of other societies but maintaining the prior culture. Third, hybridisation, where people synthesise local and other cultural phenomena to produce distinctive, hybrid forms, and lastly, transformation, where fresh ideas from other cultures are favoured and local culture may be abandoned in favour of alternative and/or hegemonic forms. These types can be influenced by cultural receptivity, socio-spatial characteristics, such as favoured meeting places and social connectivity (technologies), rituals and habitus (values), as well as internal patterns of association (structures and hierarchies).

The concept of glocalisation has been used to explain the dynamics of tourism imaginaries (Salazar, 2012), acculturation (Weedon, 2012), the development of new cultural forms (Pack *et al.*, 2012), lifestyle migrants (Torkington, 2012), product placement (Prakash & Singh, 2011) and branding 'Brand Hong Kong' (Chu, 2011). As well as export development (Lim, 2005), spatial appeal and urban regeneration (Russo & Sans, 2009), the success of theme park attractions (Matusitz, 2010) and local development responses to the global tourism industry (Milne & Ateljevic, 2001).

In the context of shopping and souvenirs, Park and Reisinger (2009) highlight cultural differences as glocal manifestations of shopping for luxury travel goods. Also, the changes that local artisans make to accommodate tourist preferences produce contemporary interpretations of traditional forms (Jena, 2010). Wilkins (2011) found differences between male and female preferences to purchase local, regional speciality and non-regional arts and crafts, and that some regional arts and crafts and local specialty products were produced to appear locally made, in response to demand generated by international tourists. The commodification of souvenirs and handicrafts is similarly influenced (Swanson & Timothy, 2012). The supply of local souvenirs is central to tourists' shopping satisfaction (Murphy *et al.*, 2011). Polsa and Fan (2011) borrow the model of Giulianotti and Robertson (2007) to describe the interrelationship between global versus local capitals in the retail industry. They note that avoidance of, or withdrawal from, global capital and retail formats emphasises highly local forms of capital, whereas complete immersion in global formats may have the effect of negating local features. However, complete withdrawal from both local and global features may create deculturation that ultimately produces totally novel forms of retail.

Ritzer (2003) suggests that, since the concepts of local, glocalisation and globalisation are about the consumption of 'something' and represent unique commodities that possess local geographic (indigenous) ties, specific to the times, that are humanised and enchanted, an oppositional companion should conceptually exist. He proposes 'grobalisation' as glocalistion's antithesis, a notion that reflects the contemporary desire by nations, corporations and organisations to 'grow' profits, power and influence throughout the world, but without substantive content, and thus result in 'nothing', such as McDonaldisation. Examples of 'glocal somethings' for Ritzer are craft barn (place), local crafts (thing), craftsperson (person) and demonstration (service). Interestingly for this chapter, examples of 'glocal nothings' for Ritzer are: souvenir (nonplace), tourist trinkets (nonthing), souvenir shop clerk (nonperson) and help-yourself (nonservice). Thus for Ritzer, souvenirs are generic, lack local ties, are not tied to a time period, dehumanised and disenchanted and not glocal.

Current trends

The global–local nexus locates in the 20th century consumer society, global mobilities and virtual technologies that emerged post-WWII (Durham & Kellner, 2006), characterised by transnational processes (Appadurai, 1990, 2001; Pieterse, 1994). Globalisation in the 21st century is very different than that of the 20th century, evidenced by pluralism, multiple modernities and different capitalisms. Today these are shaped by self-driven development agendas of the newly industrialised and agro-mineral exporting countries of the global south rather than the developed north, as well as by a swing away from 'unfettered market forces' towards state control of economic growth (Pieterse, 2012). However, problems of power and the capture of the new states by vested interests and groups within social hierarchies are evident, but not as capitalism. Transnational migrant labour is creating new epicentres of remote nationalism, cultural enclaves, social imaginaries and indigeneities that shape the economics of Florida's (2008) 'world that is not flat', but spiked by the competiveness of the rising multinationals and innovators of the global south (Pieterse, 2012). Thus the local may, in time, overtake the global or coexist alongside each other, perhaps as simultaneously glocal hybrid forms that affirm difference as well as similarity, underpinned by territorial versus translocal assumptions about culture, such as static versus fluid cultural relations (Pieterse, 1994). The reality however, is that all cultures are hybrid and are of 'place'(Mitchell, 1997).

Such global trends are abstract, expansive, large and dependent on the inventiveness of capitalism and the movements of money, commodities and people (prompted by work, education, lifestyle, ecological or political displacements, etc.) that penetrate and transform the local, but assume powerlessness at the local level (Gibson-Graham, 2008). However, beyond the forces of localisation and subversion noted above by Appadurai (2001) are community-based economies that defend traditions, use in-situ labour and possess the strength of bounded identities (Gibson-Graham, 2008). They are diverse non-capitalist spaces, such as the informal household economy, collectives, independent producers and barter networks (Gibson-Graham, 2006), that exist outside of the formal wage economy, challenging capitalism by developing alternatives to waged work and aiming to foster community wellbeing and resilience (Gibson-Graham & Roelvink, 2010).

Within this range, labour is performed and remunerated, appropriated and distributed as waged and salaried labour within capitalist forms of enterprise, but alternatively also as paid and unpaid labour in other forms of enterprise for whom private accumulation of surplus is not, or not the only, core business (Gibson-Graham, 2006). In this sphere, transformative work is undertaken by research collectives to assist communities to conceive and operationalise diverse yet sustainable economies beyond the mainstream (Cameron & Gibson-Graham, 2003; Cave et al., 2012; Gibson-Graham, 2003; Gibson-Graham & Roelvink, 2009, 2010). Examples of alternate economies in tourism include home exchange, gift-giving, voluntourism and woofing (Mosedale, 2011). Action research with nascent Pacific migrants identified a range of enterprise, labour and ceremonial alternatives that operate outside, but interface with, the formal tourism economy. These are offered by families, school, micro-enterprise and church agencies, as encounters with tourism visitors at minimal or no charge, and motivated by anticipation of future benefits to the family rather than the individual through extended networks, acquisition of knowledge and reciprocal 'returns' of hard and soft goods, as well as currency (Cave, 2009).

The second-hand clothing market in the Philippines and in Tonga (Addo & Besnier, 2008; Milgram, 2004) are examples of the reconfiguration of transnational commodities activated by women, working outside state control and creating alternative economies that cut through capitalist processes and managed using cultural practices. These are originally micro-enterprises in the informal sector, but the women expand from this base to make investments in the formal sector (Milgram, 2004), which are global networks, thus allowing 'waste' to be become a new form of resource (Lane et al., 2009). Household practices, such as subsistence activities (fishing, horticulture), arts and craft, are not driven by austerity as much as continuities of cultural tradition (Smith, 2002), as are 'do-it-yourself' repairs, construction using 'found' materials (reducing purchases of food, furnishings and furniture). These also contribute to the notion of diverse economies.

Diverse economies may not be solely located in one place, but can be linked across a wide network of producers and consumers, on the trade of locally produced crafts, goods and services that emphasise quality, authenticity, natural materials and independence from globally oriented organisations, and seen in mature economies such as Italy and Japan (McCauley, 2012). Women and men play key roles in the production of gendered handicrafts in the informal economy as surplus to cultural exchange and made for intentionally for sale, but in different forms if destined to be sold within or outside a cultural community (Cave, 2009). Swanson and Timothy identify that women encounter disempowerment and subjugation to male-dominated market economies, but that family, home and community factors facilitate their empowerment and enhance growth (Swanson & Timothy, 2012).

In the first generation of migration, migrant entrepreneurs operate largely outside the formal economy, focusing on ethnic business strategies, ethnic markets and customers, and oriented towards ethnic products. But over time, the alternative economy gradually becomes a distinctive portion of the local economy, especially in large cities (Nijkamp *et al.*, 2010). However, it may be that the pervasiveness of consumerism inevitably links the informal economy to capitalism via outsourcing and other networks (Rutherford, 2010). Yet this is not a nihilistic view. The informal economy has been called 'hidden', which implies pejorative negative view. But the informal economy is vibrant and active, achieving goals that are not solely income-based (Williams *et al.*, 2010). These vary, in urban contexts by affluence and a propensity to 'trade off the books' (unregistered) and to use methods that do not always adhere to the rules. Deprived populations are considerably more enterprising than those in affluent areas and more likely to work outside the formal economy and spin-off businesses from serious leisure (hobbies) and personal interest. Whereas, affluent populations are more likely to spin-off enterprises from formal employment (Williams & Nadin, 2010). However, both can be seen as incubators for business potential, transition to accessing wealth in the formal economy or conducted by choice by people for whom this way of working is satisfying and meets cultural, personal and identity goals.

The informal economy has, in the past, been perceived as operating in marginal populations, but it is now estimated that nearly two-thirds of the global working population are participants, but vary in permanency of engagement and level of participation (Williams *et al.*, 2010). Further, that non-market activity is persistent and growing in resistance to marketism and the market sphere is diminishing (Williams, 2004). And, the market economy can be enmeshed in, and transformed by, indigenous socio-cultural practices, producing a locally inflected version that meets local place-based indigenous goals (Curry, 2003). Thus are active agents of change who adopt and adapt global processes and reconfigure them as hybrid global–local models, but struggle to uphold community expectations and compete with other destinations without broader (state) support structures and external networks (Trau, 2012). Shopping can be construed as part of relationships of care, particularly within families (Lane *et al.*, 2009) and thus part of the social glue that holds communities together, whether in mainstream centres or on peripheries.

In the next section we ask if societal marginality and or physical distance alters the nature and dynamics of souvenir creation, production, exchange and acquisition? What happens in terms of participation in the souvenir industry?

Margins and Peripheries

We situate this book in cultural, geographic, societal and economic margins of the globe. A glocality consists of 'cultural and spatial configurations that connect places with each other to create regional spaces and regional worlds' (Escobar, 2001: 166). Glocalities are more than resistant spaces but boundaries where everyday practices are contested, mutually produced and agreed by allies/opponents (Davis, 2010). The emerging field of critical tourism addresses such issues in terms of relativities of immobility, marginality and impoverishment between tourists and hosts (Gale, 2008).

In a cultural sense, in many cases communities at the margins are self-defined. De Beauvoir (1963/1968) suggests that intentional communities simultaneously marginalise and strengthen themselves through symbiotic relationships to the mainstream, using silences and other human dynamics such as violence 'since it is only in violence that the oppressed can obtain their human status'. Alternative ways of knowing the world are seen in the ordering and meaning of 'group life' (Lichterman, 1998). There are many examples of religious intentional communities of faith – Amish, Buddhist, Hare Krisha, Jonesville Guyana, Waco and many more, such as communes. These can be defined by the attitudes and perceptions of mainstream populations, by lack of access to society's benefits, by simply choosing to live in isolated areas for lifestyle reasons, marginalised by politics, religion or lifestyle choice or by positive strategies to strengthen ethnic hegemony. The communities often share utopian ideals and coalesce around charismatic individuals and self-assured 'special' characteristics. Otherness and attributes of difference are often associated societies that are distant somehow from mainstream western thought and have been influenced by the development of relations between coloniser and colonised (Lester, 1998). The liminality of immersion in self-othered cultures is integral to their touristic experiences (Graburn, 1977).

Some groups operate as resistant communities, oppositional to the culture of the colonial and global west (Bhabha, 1990). Examples are 'identity movements' who affirm difference along identity lines, such as sexuality, gender, race or ethnicity. These include activist settings and self-defined groups who utilise and share 'practices' (Bourdieu, 1977, 1990), engage in 'civic practices' (Eliasoph, 1996) and share 'cultures of commitment' (Lichterman, 1998) or perspectives that serve to maintain cohesion and distinctiveness from the mainstream, as well as become divisors between them. However, the 'excluded middle', the liminal subaltern figures who slip between two dominant antithetical categories as described by Hegel, cited in Young (2001) should also be included.

The tourism industry is also a powerful agent of marginalisation. Tourism is the business of otherness par excellence (Hollinshead, 1998) reflexively producing the other and the tourist through mutual gaze (Maoz, 2006) as differently consumable raced, classed, gendered and sexualised subjects (Law, 2012). These contribute to choices that communities make to preserve and leverage distinctiveness in the supply of tourism products, as well as encouraging demand for a wide range of distinctive experiences and destinations. However locals and non-tourist voices are repeatedly absent in media such as travel guides that consistently overwrite local difference with a newly formulated Orientalism (Bennett, 2008). Further, tourism can place communities in direct resistance with resort development organisations to protect indigenous lands and livelihoods that have implications for the process of place-making, power and counter power (Drapeau, 2010).

Physical marginalisation of communities in peripheries can result from distances between centres, natural landscape isolation or catastrophic events (Parminter & Perkins, 1996). Ecological marginalisation occurs as ecosystems are transformed from a self-sustaining natural resource base to unproductive damaged environments as by-products of industrial or agricultural processes (Kousis, 1998). In rural areas, economic marginalisation results from changing global demands for natural resources, out-migration from rural to urban areas, as well as political and technological change, and tourism is often seen as a way to reverse the trend (Briedenhann & Wickens, 2004). Economic marginalisation is rooted in inequalities of human capital, spatial inequality and the structure of the economy, and can be characterised by underdevelopment, poverty and limitations placed on access because of structural disconnection (Philip, 2010).

Remedies to marginalisation can take the form of community developed projects, collectives and enterprise development (Cave et al., 2007; Marks, 2001; McKercher, 2001; Murphy, 1985; von Hippel, 2005) for social and economic redress. Other remedies are participative 'third sector' interventions by not-for-profit organisations, public sector policy and social programmes or perhaps community enterprise (Zapata et al., 2011). New internet-enabled technologies and decreasing cost make it possible to promote territories with strong cultural identities and enable communities to leap over other sources of disadvantage (Bordoni, 2011). Yet marginality as a positive force may contain inherent contradictions and paradox. The aspiration to establish a competitive position based on control of cultural resource (Fiol, 1991) requires the adoption of complicit sub-cultures that commodify, appropriate and alter the nature of cultural identity, but can further isolate those communities.

The sense 'being at the margin' is a self-generating (autopoetic) process of self-referent thought and identity, defined by a community as identity strategies constructed about themselves and the others in that system (Arvidsson, 1997). The concept of autopoesis is allied to habitus that refers to dynamically created values and structures common to marginalised selves. These constantly evolve and are both open to others in the external environment and yet closed, dependent upon the nature of the internal environments' relationships producing parallel strategies for cultural identity and hegemony, nurtured through grass-roots practices.

A geographic periphery is defined by Boller *et al.* (2010) as a duality between core and peripheral regions on the basis of economics, geography and destination characteristics of interest to travellers. Core areas are main tourism entry points into the country at which structures, populations and activities of modern civilisation are concentrated (Boller *et al.*, 2010). However, a peripheral tourism destination is a region that possesses unique attractions and novel experiences not available in the core (Iles & Prideaux, 2011). Thus a notion of duality is difficult to maintain since 'peripheries' imply connectedness, not only with a 'core' but also with 'remote' places beyond a periphery, at the same time as implying 'disconnectedness' (Schmallegger *et al.*, 2010) so that peripheries are nuanced by distance and socio-economic responses.

Peripheral or remote areas become depleted of labour and capital resources diverted to support the core, yet their remoteness and socio-economic difference frequently coincide with natural, historical and cultural landscapes, preserved intact to provide tourism potential (Wieckowaski, 2010). Peripheries can capture the imagination as mysterious, unknown places (Yang, 2011) and provide places of adventure that are dystopic, to be avoided, yet offer the frisson of risk that is attractive to some (Cave & Ryan, 2007) and liminal experiences of adventure in remote landscapes (Andrews, 2012).

Synergies can develop in the collaborative marketing of remote places and destinations at the peripheries of tourism flow to position their tourism products in a common market (Thorhallur & Gunnar, 2012). However, tourism peripheries also experience the paradox that tourism development depletes the experiential qualities that attracted tourists in the first instance (Boller *et al.*, 2010) and require careful planning to weigh up the benefits and the costs (Kauppila *et al.*, 2009) and sustainability of the peripheral economy, which is the focus of the majority of most research.

Further, tourism experiences and handicraft production are synergistic and symbiotic. Tourists provide a market for the continuance of traditional art forms. Yet handicraft and souvenir businesses are critical forces in the survival of remote communities (Kauppila *et al.*, 2009), the development of tourism in peripheral regions (Fonseca & Ramos, 2012), as well as income generators for indigenous communities (Lacher & Nepal, 2010; Pratt *et al.*, 2012). Souvenirs also reflect rural identity (Rogerson & Rogerson, 2012) and destination distinctiveness (Prideaux & McNamara, 2012).

The concepts of glocal, societal and physical peripheries, and the animating processes of habitus, otherness and autopoeisis are important to this book because they describe a dynamic, complex system in which the tourism industry sits, spanning the local informal economy, a glocal (hybrid) interface of informal/formal, and more globalised formal economies. Figure 1.1 proposes a synthesis of glocalisation, cultural and otherness theory to model the dynamics of choices (adopt/avoid) by communities and individuals at social, cultural, physical, ecological, economic, political margins (peripheries).

Conceptually, the local informal household economy produces identity, predicated on choices by communities as to the adoption or avoidance of the ideas of others. The glocal hybrid interface is notionally where touristic encounters and tourism transactions locate (Cave, 2005). As noted earlier in the chapter, relativisation, accommodation, hybridisation and transformation

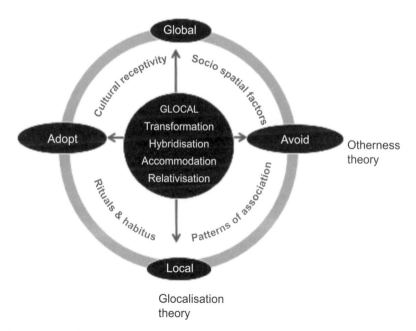

Figure 1.1 Framing tourism and souvenirs at the margins (after Cave (2005) and Giulianotti & Robertson (2007))

are hybridising cultural glocalisation practices that respond, resist and inter-pret external influences, relative to local conditions. They are dependent on the factors of cultural receptivity, social spatial features, rituals and habitus and patterns of association (Giulianotti & Robertson, 2007). For example, where people meet, within which social hierarchies, their values and recep-tiveness to others. Such processes produce differing emphases on localised versus globalised cultural practices. The formal market economy is increas-ingly globalised (perhaps grobalised) and commercial.

This framework is important to the book because each chapter of the volume provides differing lenses on this dynamic system. And further, each author illustrates separate yet interconnected cultural glocalisation practices of cultural receptivity, socio-spatial characteristics, external connectivity and internal patterns of association within global and local social worlds.

Chapter Structure

This first chapter, written by the editors Jenny Cave, Tom Baum and Lee Jolliffe, theorises souvenirs in terms of glocal perspectives, marginality and tourism peripheries. These concepts are further addressed by the author con-tributors to this volume under three thematic sections: Part 1 – Theorising Experience and Behaviour, Part 2 – Theorising Place and Identity, and Part 3 – Glocal Case Studies in Sustainable Tourism.

Part 1 – Theorising Experience and Behaviour

Noga Collins-Kreiner and Yael Zins (Chapter 2) explore the tangible, mate-rial importance of souvenirs, tourist owners' attitudes toward objects and their glocal meanings in peoples' homes. They theorise time-dependent behaviour signification and changing attributions of meaning and importance. Adding to the literature on souvenir acquisition they argue for a decay of acquisitiveness as tourists become more familiar with a destination and that marginalisation of the objects takes place in the home environment over time. Yet dynamic shifts occur in where the practical objects can become more important than the 'official' souvenir, thus they recommend expanding the notion of 'souve-nir' to include practical/functional objects purchased abroad.

Hugh Wilkins's chapter (Chapter 3) theorises linkage between souvenir purchase and the construction of consumer identity, specifically the intan-gible consequences of gathering of souvenirs as mementos of experience, either for consumption by others or as means of prolonging one's own con-sumption experience. The analysis extends prior research by applying the concept of congruence in souvenir consumption, relating souvenir purchase behaviour to aspirational self-image and status, as well as to an evaluation of the relationships between the attributes of souvenirs (e.g. exclusive, aesthetic, authentic) and purpose, such as gift, aide memoire and evidence.

Richard Robinson's chapter (Chapter 4) does not look at the consumption of tangible touristic artefacts, nor intangible sensory experiences. Rather it contributes theorisation and evidence of a previously ignored category of souvenirs that relate to the supply of food consumption and foodservices to tourism experiences – occupational artefacts. This study explores the ways that chefs acquire and valorise occupational artefacts that express an intrinsic sense of global community and culture, yet localise their place within it. Dimensions identified in this pioneering study are the nature of community, authenticity, identity and belonging (in time and place), marginality and deviance, mobility and the human resource challenges they encounter.

Part 2 – Theorising Place and Identity

Kirsten Swanson contributes an insightful examination of objective, subjective and constructive authenticity. In Chapter 5 she analyses the effects that acculturation and colonisation have had on the creation and performance of culture and place. This establishes that the Apache, Hopi, Navajo, Pueblo and Zuni souvenirs that most represent 'objectively authentic' in modern times are in fact 'constructed authentic' mainstream representations of the ancient indigenous cultures – which marginalise the original, authentic and the real. The author argues that tourist perceptions of place and people are central to reassertions of culturally sustainable indigenous production and authority over authenticity.

Russell Staiff and Robyn Bushell (Chapter 6) conduct an ethnography of commodities to look at the social/cultural processes that produce and consume souvenirs. The authors argue that Laotian-carved Buddha sculptures, are marginalised by the culture of their producers since they are contemporary appropriations of ancient forms, deliberately 'aged' to enhance their appeal for sale as souvenirs. They further suggest that the Western, culturally produced hierarchy of material objects that places art/design at the apex and 'kitsch'/souvenirs at the bottom, marginalises souvenirs within tourism's material culture and that the association with longing and nostalgia borne by the objects further distances their worth. However, the artistic value and place-based uniqueness of these objects have global ramifications as collectibles and are economically central to the material culture of Luang Prabang, a city produced by centuries of local–global economic flows.

Jenny Cave and Dorina Buda adopt a poststructural critical turn in Chapter 7 to theorise place-associated meanings and identity of the supply of souvenirs by retailers as glocal encounters with difference. This study utilises an innovative model for understanding tourism interactions as non-cash encounters, as well as cash-based transactions in the supply-side of tourism (hosts). This forms a base from which to retheorise the implications of authenticity to suppliers across the local informal economy, as a glocal hybrid interface of informal/formal, and more globalised formal economy. The study, undertaken in Coastal Bay of Plenty, New Zealand, encompasses a range of formal and informal tourism retail formats (some indigenous), microenterprise producers and publicly funded operations and asks, to what extent do souvenir retailers play a role in the portrayal of place and identity to consumers, and their perceptions of souvenir authenticity and commodification?

Part 3 – Glocal Case Studies in Sustainable Tourism

Atsuko Hashimoto and David Telfer (Chapter 8) provide a study of the role that souvenirs play in the empowerment of elderly women in a peripheral region of rural Japan, whose lack of external contact produces high cultural authenticity. The authors look at the challenges faced by local women whose production of innovative souvenirs for sale conflicts with traditional farming, tourism seasonality and support from family members. Yet benefits are gained in social networking, socialisation and an income supplement that offsets isolation and meagre livelihoods. The setting for the analysis is Green Tourism, which includes agritourism-based hotels, farmers' markets, ecomuseums, bed and breakfasts, and hot-spring spas, encouraged by government initiatives as economic rejuvenation.

Laurie Murphy, Gianna Moscardo and Pierre Benckendorff (Chapter 9) examine the challenges and opportunities experienced by small rural villages which offer tourist shopping (Tourist Shopping Villages) as a way to support local production of arts, crafts, and specialist food and beverage, on the periphery of the urban centres of Adelaide and Brisbane, Australia. In particular they look at the strength of place-specific connections between tourist experience and actual purchase behaviours. The authors identify a strong link between an enjoyable tourist experience in a Tourist Shopping Village and the likelihood and extent of tourists purchasing local souvenirs. Further, that village atmosphere, convenience, variety, experiential activities and purchase opportunity are factors that can provide more income for local communities.

Geoffrey Lacher and Susan Slocum in Chapter 10 compare remote destinations in northern Tanzania and Thailand to explore the constraints and complexities that peripheral communities face when they attempt to enter the souvenir market. They debate the ambiguities of souvenir production as a supplemental source of income for peoples with few alternatives in rural peripheries, or a means of their exploitation by external suppliers of mass-produced items. The chapter identifies effective remedies, such as distinguishing locally made items from imported goods, forming cooperatives and providing organised assistance with business education and market intelligence. These increase the amount of local goods sold and reduce leakage, but balances these with the realties that are encountered on a daily basis by these communities.

Huong Bui, and Lee Jolliffe (Chapter 11) examine the issue of transitioning from production to attraction in terms of the creation of place-based souvenirs in developing rather than developed countries. The authors profile two handicraft villages, one on the periphery of a historic city, the other on the edge of an ancient town in Vietnam, exploring issues of authenticity and commercialisation. The authors examine how craft-producing villages evolve over time from sales of local goods, to specialised retailing of imported goods, to 'ribbon' concentrations of craft shops, workshops and services that eventually become tourist attractions in their own right. The transition process is heavily influenced by proximity to tourist-generating centres, depth of handicraft tradition, stakeholder involvement (community, private sector, public sector, non-governmental organisations and global tourism) and the attractiveness of the village setting.

Hilary du Cros (Chapter 12) addresses several gaps in the literature: Asian domestic tourism, domestic tourism souveniring behaviours, the influence of modernity and global marketing practices, and the role that food plays as a souvenir. It looks at the meaning of cultural heritage mementos purchased by Asian tourists to Macau, and the motives behind that action. The author identifies developmental and spatial indicators regarding the ways that souvenir shops at a World Heritage site have responded to a perceived market shift towards domestic tourism from Mainland China. Du Cros asks whether that shift is reflected in the material culture of tourism, particularly in the entrepreneurial innovation of baked food that are iconic to Chinese culture, as souvenirs.

Conclusions

In the final chapter, Lee Jolliffe, Jenny Cave and Tom Baum (Chapter 13 – Lessons in tourism and souvenirs from the margins: Glocal perspectives) offer comparisons of the issues raised by chapter authors and proposes new theorisation of the field of tourism and souvenirs – from the margins. This chapter builds on current literature, looking at the material from a glocal lens in terms of synergies between the global and local. In this final chapter we synthesise research themes common to the chapters, outline research gaps from the evidence identified by the authors and conclude with an analysis of academic frontiers and the contributions to new knowledge.

This volume develops new theory on souvenirs as tangible *triggers* of tourism experience, looks at tourism product as material culture and addresses issues of sustainability and community development in the peripheries. Further, it explores the role that tourism plays in economic development, commenting on social and cultural preservation for communities and entrepreneurs located on island, urban and rural peripheries around the globe and on the margins of society.

Our call for proposals produced a positive response from established and emerging researchers demonstrating a strong research interest in the topic, from both academics and practitioners working in a number of areas. Also this work evidences growing theoretical integrity within this subfield of the tourism research agenda, and includes the work of anthropologists, museologists, cultural studies scholars, tourism and hospitality management academics, geographers and sociologists.

The book will be useful to community development projects, producers and suppliers, and micro-entrepreneurs for touristic projects, as well as small- and medium-sized entrepreneurs in the formal industry, attractions developers, marketers and destination management organisations in the experience economy. Academics, practitioners and policy makers will find this volume a valuable reference book, as will senior and graduate student researchers. Academic teachers will be able to use the volume as a textbook, or chapters as class resources and case examples for anthropology, cultural studies, critical theory, economic geography and tourism and hospitality management.

References

Adams, K. (1997) Ethnic tourism and the regeneration of tradition in Tana Toraja (Sulawesi, Indonesia). *Ethnology* 36 (4), 1–10.

Addo, P.A. and Besnier, N. (2008) When Gifts become commodities: Pawnshops, valuables, and shame in Tonga and the Tongan Diaspora. *Journal of the Royal Anthropological Institute* 14, 39–49.

Anderson, A. and Littrell, M.A. (1995) Souvenir-purchase behavior of women tourists. *Annals of Tourism Research* 22 (2), 328–348.

Andrews, H. (2012) Introduction. *Journal of Tourism Consumption and Practice* 4 (1), 1–4.

Appadurai, A. (1990) Disjuncture and difference in the global cultural economy. *Public Culture* 2 (2), 1–24.

Appadurai, A. (2001) Globalization, Anthropology of. In N.S. Smelser and P.B. Baltes (eds) *International Encyclopedia of the Social & Behavioral Sciences* (pp. 6266–6271). Oxford: Elsevier.

Arvidsson, A. (1997) Reconstructing the public sphere: AST and the observation of post-modernity. *Kybernetes* 26 (6/7), 661–673.

Asplet, M. and Cooper, M. (2000) Cultural designs in New Zealand souvenir clothing: The question of authenticity. *Tourism Management* 21 (3), 307–312.

Bennett, R.J. (2008) Entering the global margin: Setting the 'other' scene in independent travel. In P.M. Burns and M. Novelli (eds) *Tourism and Mobilities: Local–Global Connections* (pp. 133–145). CAB International.

Bhabha, H.K. (1990) The other question. In R. Ferguson, M. Gever, T. Minh-ha and C. West (eds) *Out There: Marginalization and Contemporary Cultures* (pp. 71–80). New York: MIT Press.

Boller, F., Hunziker, M., Conedera, M., Elsasser, H. and Krebs, P. (2010) Fascinating remoteness: The dilemma of hiking tourism development in peripheral mountain areas. *Mountain Research and Development* 30 (4), 320–331.

Bordoni, L. (2011) Technologies to support cultural tourism for Latin Latium. *Journal of Hospitality and Tourism Technology* 2 (2), 96–104.

Bourdieu, P. (1977) *Outline of a Theory of Practice.* Cambridge: Cambridge University Press.

Bourdieu, P. (1990) *Structures, Habitus, Practices* (R. Nice, Trans.). Cambridge: Polity Press.

Boynton, L. (1986) The effect of tourism on amish quilting design. *Annals of Tourism Research* 13 (3), 451–465.

Briedenhann, J. and Wickens, E. (2004) Tourism routes as a tool for the economic development of rural areas—vibrant hope or impossible dream? *Tourism Management* 25 (1), 71–79.

Cameron, J. and Gibson-Graham, J.K. (2003) Feminising the economy: Metaphors, strategies, politics. *Gender, Place & Culture* 10 (2), 145–157.

Cave, J. (2005) Conceptualising 'otherness' as a management framework for tourism enterprise. In C. Ryan and M. Aicken (eds) *Indigenous Tourism – the Commodification and Management of Culture* (pp. 315–334). Oxford: Pergamon.

Cave, J. (2009) Embedded entrepreneurs: Nascent tourism and diasporan Pacific Islanders. In J. Carlsen, M. Hughes, K. Holmes and R. Jones (eds) *See Change: Tourism and Hospitality in a Dynamic World* (pp. 1–8). Fremantle, Australia.

Cave, J., Johnston, L., Morrison, C-A. and Underhill-Sem, Y. (2012) Community-university collaborations: Creating hybrid research and collective identities. *Kotuitui: New Zealand Journal of Social Sciences Online* 7 (1), 37–50.

Cave, J., Jolliffe, L. and De Coteau, D. (2012) Mementos of place: Souvenir purchases at the Bridgetown Cruise Terminal in Barbados. *Tourism, Culture & Communication* 12 (1), 39–50.

Cave, J. and Ryan, C. (2007) Gender in backpacking and adventure tourism. *Advances in Culture, Tourism and Hospitality Research* 1, 183–214.

Cave, J., Ryan, C. and Panakera, C. (2007) Cultural tourism product: Pacific Island migrant perspectives in New Zealand. *Journal of Travel Research* 45, 435–443.

Chu, S.Y-W. (2011) Brand Hong Kong: Asia's world city as method? *Visual Anthropology* 24 (1), 46–58.

Cohen, A. (1983) The dynamics of commercialised arts: The Mao and Yao of Northern Thailand. *Journal of National Research Council of Thailand* 15 (1, Pt. II.), 1–34.

Cohen, A. (1985) *The Symbolic Construction of Community.* Sussex: Ellis Horwood.

Cohen, E. (1988) Authenticity and commoditization in tourism. *Annals of Tourism Research* 15, 371–386.

Cohen, E. (1993) The heterogenization of a tourist art. *Annals of Tourism Research* 20 (1), 138–163.

Cohen, E. (1993b) Investigating tourist arts. *Annals of Tourism Research* 20 (1), 1–8.

Cohen, E. (2001) Ethnic tourism in Southeast Asia. In C-B. Tan, C-H. Sidney and Y-H. Cheung (eds) *Tourism, Anthroplogy and China* (pp. 27–53). Singapore: White Lotus Press.

Curry, G.N. (2003) Moving beyond postdevelopment: Facilitating indigenous alternatives for "development". *Economic Geography* 79 (4), 405–423.

Davis, L. (2010) *Alliances: Re/Envisioning Indigenous-non-Indigenous Relationships*. Toronto: University of Toronto Press.

de Beauvoir, S. (1963/1968) *Force of Circumstance* (R. Howard, Trans.). Harmondsworth: Penguin.

Deitch, L. (1989) The impact of tourism upon the arts and crafts of the Indians of the Southwestern United States. In V. L. Smith (ed.) *Hosts and guest: The Anthropology of Tourism* (Vol. 173–184). Philadelphia: University of Pennsylvania.

Dirlik, A. (1999) Place-based imagination: Globalism and the politics of place. *Review* XXII (2), 151–187.

Doorne, S., Ateljevic, I. and Bai, Z. (2003) Representing identities through tourism: Encounters of ethnic minorities in Dali, Yunnan Province, People's Republic of China. *International Journal of Tourism Research* 5 (1), 1–11.

Drapeau, T. (2010) A glocality in the making: Learning from the experience of resistance of the Secwepemc Watershed Committee against Sun Peaks Resort, British Columbia. In L. Davis (ed.) *Alliances: Re/Envisioning Indigenous-non-Indigenous Relationships* (pp. 213–233). Toronto: University of Toronto Press.

Durham, M.G. and Kellner, D.M. (2006) Adventures in media and cultural studies: Introducing the KeyWorks. In D.M. Kellner and M.G. Durham (eds) *Media and Cultural Studies: Keyworks* (Revised ed.). Malden, Oxford and Victoria: Blackwell Publishing.

Eliasoph, N. (1996) Making a fragile public: A talk-centered study of citizenship and power. *Sociological Theory* 14, 262–289.

Escobar, A. (2001) Culture sits in places: Reflections on globalism and subaltern strategies of localization. *Political Geography* 20, 139–174.

Ezarik, M. (2003) Appreciating the beauty of cultural diversity. *Current Health 2* 29 (6), 25–28.

Fairhurst, A., Costello, C. and Holmes, A.F. (2007) An examination of shopping behaviour of visitors to Tennessee according to tourist typologies. *Journal of Vacation Marketing* 13 (4), 311–320.

Fonseca, F.P. and Ramos, R.A.R. (2012) Heritage tourism in peripheral areas: Development strategies and constraints. *Tourism Geographies: An International Journal of Tourism Space, Place and Environment* 14 (3), 467–493.

Fiol, C.M. (1991) Managing culture as a competitive resource: An identity-based view of sustainable competitive advantage. *Journal of Management* 17, 191–211.

Florida, R. (2008) *Who's Your City?* New York: Basic Books.

Forsyth, T. (1995) Tourism and agricultural development in Thailand. *Annals of Tourism Research* 22 (4), 877–900.

Fuller, D.J.B. and Cummings, E. (2005) Ecotourism and indigenous micro-enterprise formation in northern Australia opportunities and constraints. *Tourism Management* 26 (6), 891–904.

Gale, T. (2008) The end of tourism, or endings in tourism? In P.M. Burns and M. Novelli (eds) *Tourism and Mobilities: Local–Global Connections* (pp. 1–14). CAB International.

Gibson-Graham, J.K. (2002) Beyond global vs. local: Economic politics outside the binary frame. In A. Herod and M. Wright (eds) *Geographies of Power: Placing Scale* (pp. 24–60). Oxford: Blackwell Publishers.

Gibson-Graham, J.K. (2003) Enabling ethical economies: Cooperativism and class. *Critical Sociology* 29 (2), 123–161.

Gibson-Graham, J.K. (2006) *The End of Capitalism (As We Knew It): A Feminist Critique of Political Economy* (1st ed.). Minneapolis, US: University of Minnesota Press.

Gibson-Graham, J.K. (2008) Diverse economies: Performative practices for 'other worlds'. *Progress in Human Geography* 32, 613–632.

Gibson-Graham, J.K. and Roelvink, G. (2009) Social innovation for community economies. In D. MacCallum, F. Moulaert, J. Hillier and S. Haddock (eds) *Social Innovation and Territorial Development* (pp. 25–37). Farnham, UK: Ashgate.

Gibson-Graham, J.K. and Roelvink, G. (2010) The nitty gritty of creating alternative economies. *Social Alternatives* 30 (1), 29–33.

Giulianotti, R. and Robertson, R. (2007) Glocalization, globalization and migration: The case of scottish football supporters in North America. *International Sociology* 21 (2), 171–198.

Gordon, B. (1986) The souvenir: Messenger of the extraordinary. *Journal of Popular Culture* 20 (3), 135–151.

Graburn, N. (1976) *Ethnic and Tourist Arts. Cultural Expressions from the Fourth World.* Berkeley: University of California Press.

Graburn, N. (1977) Tourism: the sacred journey. In V. Smith (ed.) *Hosts and Guests: The Anthropology of Tourism* (pp. 17–31). Philadelphia, Pennsylvania: University of Pennsylvania Press.

Grünewald, R. (2006) Pataxó tourism art and cultural authenticity. In M. Smith and M. Robinson (eds) *Cultural Tourism in a Changing World: Politics, Participation and (Re) Presentation* (pp. 203–214). Clevedon: Channel View Publications.

Haggard, L. and Williams, D. (1992) Identity affirmation through leisure activities: Leisure symbols of the self. *Journal of Leisure Research* 24 (1), 1–18.

Haldrup, M. (2009) Local-global. In K. Rob and T. Nigel (eds) *International Encyclopedia of Human Geography* (pp. 245–255). Oxford, England: Elsevier.

Hamel, N. (2001) Coordonner l'artisanat et le tourisme, ou comment mettre en valeur le visage pittoresque du Québec (1915–1960). *Histoire Sociale* 34 (67), 97–114.

Hashimoto, A. and Telfer, D.J. (2008) Geographical representations embedded within souvenirs in Niagara: The case of geographically displaced authenticity. *Tourism Geographies* 9 (2), 191–217.

Healy, R. (1994) Tourist merchandise as a means of generating local benefits from eco-tourism. *Journal of Sustainable Tourism* 2 (3), 137–151.

Hitchcock, M. and Teague, K. (2000) *Souvenirs: The Material Culture of Tourism.* Aldershot: Asghate Publishing Ltd.

Holder, J. (1989) Tourism and the future of Caribbean handicraft. *Tourism Management* 10 (4), 310–314.

Hollinshead, K. (1998) Tourism, hybridity and ambiguity: The relevance of Bhabha's 'Third Space' cultures. *Journal of Leisure Research* 30 (1), 121–156.

Hollinshead, K., Ateljevic, I. and Ali, N. (2009) 'Tourism state' cultural production: The re-making of Nova Scotia. *Tourism Geographies* 11 (4), 526–545.

Hume, D. (2009) The development of tourist art and souvenirs – the arc of the boomerang: from hunting, fighting and ceremony to tourist souvenir. *International Journal of Tourism Research* 11 (1), 55–70.

Iles, A. and Prideaux, B. (2011) *The Effect on a Peripheral Destination from Changes to the Working Backpacking Market in Australia.* Paper presented at the CAUTHE National Conference, 8–11 February 2011, Adelaide.

Jena, P.K. (2010) Indian handicrafts in globalization times: An analysis of global-local dynamics. *Interdisciplinary Description of Complex Systems* 8 (2), 119–137.

Jolliffe, L. (2007) Tea and travel: Transforming the material culture of tea. In L. Jolliffe (ed.) *Tea and Tourism: Tourists, Traditions, Transformations* (pp. 23–37). Clevedon: Channel View Publications.

Jules-Rosette, B. (1984) *The Messages of Tourist Art: An African Semiotic System in Comparative Perspective.* New York: Plenum Press.

Kauppila, P., Saarinen, J. and Leinonen, R. (2009) Sustainable tourism planning and regional development in peripheries: A Nordic VIEW. *Scandinavian Journal of Hospitality and Tourism* 9 (4), 424–435.

Kim, S. and Littrell, M.A. (1999) Predicting souvenir purchase intentions. *Journal of Travel Research* 38, 153–162.

Kim, S. and Littrell, M.A. (2001) Souvenir buying intentions for self versus others. *Annals of Tourism Research* 28 (3), 638–657.

Kleymeyer, C.D. (1994) *Cultural Expression and Grassroots Development: Cases from Latin America and the Caribbean.* Boulder, CO: L. Rienner.

Kousis, M. (1998) Ecological marginalisation in rural areas: Actors, impacts, responses. *Sociologica Ruralis* 38 (1), 86–108.

Kumar, M. and Janz, T. (2010) An exploration of cultural activities of Métis in Canada. *Canadian Social Trends* 89, 63–69.

Lacher, R.G. and Nepal, S.K. (2010) From leakages to linkages: Local-level strategies for capturing tourism revenue in Northern Thailand. *Tourism Geographies* 12 (1), 77–99.

Lane, R., Horne, R. and Bicknell, J. (2009) Routes of reuse of second-hand goods in Melbourne households. *Australian Geographer* 40 (2), 151–168.

Law, P. (2012) Ethnic tourism: The spectacle of the other. *Prospect, April* Retrieved from http://prospectjournal.org/2012/04/20/ethnic-tourism-the-spectacle-of-the-other/

Lester, A. (1998) 'Otherness' and the frontiers of empire: The Eastern Cape Colony, 1806– c. 1850. *Journal of Historical Geography* 24 (1), 2–19.

Lichterman, P. (1998) What do movements mean? The value of participant-observation. *Qualitative Sociology* 21 (4), 401–418.

Lim, E.B. (2005) The Mardi Gras Boys of Singapore's English-Language Theatre. *Asian Theatre Journal*, 22, 293–309.

Lippard, L. (1999) *On the Beaten Track : Tourism, Art and Place.* New York: New Press.

Littrell, M.A., Baizerman, S., Kean, R., Gahring, S., Niemeyer, S., Reilly, R. and Stout, J. (1994) Souvenirs and tourism styles. *Journal of Travel Research* 33 (1), 3–11.

Littrell, M.A. and Anderson, L.A. (1993) What makes a craft souvenir authentic? *Annals of Tourism Research* 20 (1), 197–215.

Lowe, M.S., Williams, A.M., Shaw, G. and Cudworth, K. (2012) Self organising innovation networks, mobile knowledge carriers and diasporas: Insights from a pioneering boutique hotel chain. *Journal of Economic Geography* 12, 1113–1138.

Lury, C. (1997) The objects of travel. In C. Rojek and J. Urry (eds) *Touring Cultures: Transformations of Travel and Theory* (pp. 75–95). London, UK: Routledge.

Maoz, D. (2006) The mutual gaze. *Annals of Tourism Research* 33 (1), 221–239.

Marks, S. (2001) Back to the future: Some unintended consequences of Zambia's community-based wildlife program (ADMADE). *Africa Today* 48 (1), 121–141.

Matusitz, J. (2010) Disneyland Paris: A case analysis demonstrating how glocalization works. *Journal of Strategic Marketing* 18 (3), 223–237.

McCauley, S. (2012) Craft economies in Japan: The re-emergence of alternative economies in a no-growth context. In S. Lorek and J. Backhaus (eds) *Sustainable Consumption Transitions Series* (Vol. 1, pp. 79–88). Bregenz, Austria: Sustainable Consumption Research and Action Initiative.

McKercher, B. and du Cros, H. (2002) *Cultural Tourism: The Partnership between Tourism and Cultural Heritage Management*. New York: The Haworth Hospitality Press.

McKercher, M. (2001) Attitudes to a non-viable community-owned heritage tourist attraction. *Journal of Sustainable Tourism* 9 (1), 29–43.

Meethan, K., Anderson, A. and Miles, S. (eds) (2006) *Tourism Consumption and Representation: Narratives of Place and Self*. Wallingford: CAB International.

Miettinen, S. (2006) Raising the status of Lappish communities through tourism development. In M.K. Smith and M. Robinson (eds) *Cultural Tourism in a Changing World: Politics, Participation and (Re)presentation* (pp. 159–174). Clevedon: Channel View Publications.

Milgram, B.L. (2004) Refashioning commodities: Women and the source of second hand clothing in the Philippines. *Canadian Anthropology Society* 46 (2), 189–202.

Milne, S. and Ateljevic, I. (2001) Tourism, economic development and the global-local nexus: Theory embracing complexity. *Tourism Geographies* 3 (4), 369–393.

Mitchell, K. (1997) Different diasporas and the hype of hybridity. *Environment and Planning D: Society and Space* 15 (5), 533–553.

Moreno, J. and Littrell, M.A. (2001) Negotiating tradition: Tourism retailer in Guatemala. *Annals of Tourism Research* 28 (3), 658–685.

Mosedale, J. (2011) Diverse economies and alternative economic practices in tourism. In N. Morgan, I. Ateljevic and A. Pritchard (eds) *The Critical Turn in Tourism Studies: Creating an Academy of Hope* (pp. 194–207). London and New York: Routledge.

Murphy, L., Moscardo, G., Benckendorff, P. and Pearce, P. (2011) Evaluating tourist satisfaction with the retail experience in a typical tourist shopping village. *Journal of Retailing and Consumer Services* 18, 302–310.

Murphy, P. (1985) *Tourism – A Community Approach*. London: Routledge.

Nijkamp, P., Sahin, M. and Baycan-Levent, T. (2010) Migrant entrepreneurship and new urban economic opportunities: Identification of critical success factors by means of qualitative pattern recognition analysis. *Tijdschrift voor Economische en Sociale Geografie* 101 (4), 371–391.

Novotny, R., Moravec, I. and Abele, J. (2008) New approaches to rural development and cultural heritage preservation: Visitor Centre as a viable tool of cultural heritage presentation and microcluster in peripheral areas. Paper presented at the 4th International Scientific Conference, Jelgava, Latvia, 25–26 September 2008.

Pack, S., Eblin, M. and Walther, C. (2012) Water puppetry in the Red River Delta and beyond: Tourism and the commodification of an ancient tradition. *ASIANetwork Exchange: A Journal for Asian Studies in the Liberal Arts* 19 (2), 23–31.

Park, K.S. and Reisinger, Y. (2009) Cultural differences in shopping for luxury goods: Western, Asian, and Hispanic tourists. *Journal of Travel & Tourism Marketing* 26 (8), 762–777.

Parminter, T. and Perkins, A. (1996) The application of systems analyses to group goal setting. Paper presented at the New Zealand Agricultural Economics Society Conference, Blenheim, July 5–6 1996.

Parris, D. (1984) *Guidelines for Enhancing the Positive Socio-Cultural Impacts of Tourism in the Caribbean* (Vol. 1). Washington, DC: Organization of American States.

Perkins, M. and Morphy, H. (eds) (2006) *The Anthropology of Art: A Reader*. Malden, MA: Blackwell Publishing.

Philip, K. (2010) Inequality and economic marginalisation: How the structure of the economy impacts on opportunities on the margins. *Law, Democracy & Development* 14, 105–132.

Pieterse, J.N. (1994) Globalisation as hybridisation. *International Sociology* 9, 161–184.

Pieterse, J.N. (2012) Twenty-first century globalization: A new development era. *Forum for Development Studies* 39 (3), 367–385.

Polsa, P. and Fan, X. (2011) Globalization of local retailing: Threat or opportunity? The case of food retailing in Guilin, China. *Journal of Macromarketing* 31 (3), 291–311.

Prakash, A. and Singh, V. (2011) Glocalization in food business: Strategies of adaptation to local needs and demands. *Asian Journal of Technology & Management Research* 1 (01) 1–21.

Pratt, S., Gibson, D. and Movono, A. (2012) Tribal tourism in Fiji: An application and extension of Smith's 4Hs of indigenous tourism. *Asia Pacific Journal of Tourism Research* DOI: 10.1080/10941665.2012.717957, 1–19.

Prideaux, B. and McNamara, K.E. (2012) Turning a global crisis into a tourism opportunity: The perspective from Tuvalu. *International Journal of Tourism Research* Early View DOI: 10.1002/jtr.1883.

Reisinger, Y. and Steiner, C.J. (2006) Reconceptualizing object authenticity. *Annals of Tourism Research* 33 (1), 65–86.

Rinzin, C. (2006) *On the Middle Path: The Social Basis for Sustainable Development in Bhutan* (Vol. 352). Utrecht: Netherlands Geographical Studies.

Ritzer, G. (2003) Rethinking globalization: Glocalization/globalizaton and something/ nothing. *Sociological Theory* 21 (3), 193–209.

Rogerson, C.M. and Rogerson, J.M. (2011) Craft routes for developing craft business in South Africa: Is it a good practice or limited policy option? *African Journal of Business Management* 5 (30), 11736–11748.

Russo, A.P. and Sans, A.A. (2009) Student communities and landscapes of creativity: How Venice – 'The world's most touristed city' – is changing. *European Urban and Regional Studies* 16 (2), 161–175.

Rutherford, T.D. (2010) Labor geography. In R. Kitchen and N. Thrift (eds) *International Encyclopedia of Human Geography* (pp. 72–78). Amsterdam: Elsevier Press.

Salazar, N.B. (2012) Tourism imaginaries: A conceptual approach. *Annals of Tourism Research* 39 (2), 863–882.

Schmallegger, D., Carson, D. and Tremblay, P. (2010) The economic geography of remote tourism: The problem of connection seeking. *Tourism Analysis* 15 (1), 125–137.

Shamir, B. (1992) Some correlates of leisure identity salience: Three exploratory studies. *Journal of Leisure Research* 24 (4), 310–323.

Sharpley, R. (1994) Tourism and authenticity. In R. Sharpley (ed.) *Tourism, Tourists and Society* (pp. 127–162). Huntingdon, UK: Elm Publications.

Smith, A. (2002) Culture/economy and spaces of economic practice: Positioning households in post-communism. *Transactions of the Institute of British Geographers* 27 (2), 232–250.

Swanson, K.K. and Timothy, D.J. (2012) Souvenirs: Icons of meaning, commercialization and commoditization. *Tourism Management* 33, 489–499.

Thorhallur, G. and Gunnar, M. (2012) North Atlantic island destinations in tourists' minds. *International Journal of Culture, Tourism and Hospitality Research* 6 (2), 114–123.

Timothy, D.J. (2005) *Shopping Tourism, Retailing and Leisure*. Clevedon: Channel View Publications.

Timothy, D.J. and Wall, G. (1997) Selling to tourists: Indonesian street vendors. *Annals of Tourism Research* 24 (2), 322–340.

Torkington, K. (2012) Place and lifestyle migration: The discursive construction of 'glocal' place-identity. *Mobilities* 7 (1), 71–92.

Tosun, C. (2000) Limits to community participation in the tourism development process in developing countries. *Tourism Management* 21 (6), 613–633.

Townsend, P. and Cave, J. (2004) Working with the women – an evaluation of the impact of a papermaking project on a rural Fijian village. *Deckle Edge, Newsletter of Papermakers of Victoria Inc, Australia* 16 (3), 10–12.

Trau, A.M. (2012) Beyond pro-poor tourism: (Re)interpreting tourism-based approaches to poverty alleviation in Vanuatu. *Tourism Planning & Development* 9 (2), 149–164.

Turner, L. and Reisinger, Y. (2001) Shopping satisfaction for domestic tourists. *Journal of Retailing and Consumer Services* 8, 15–27.

von Hippel, E. (2005) *Democratizing innovation*. Cambridge, MA: MIT Press.

Vorlaufer, K. (1999) Tourism and cultural change in Bali [Tourismus und kulturwandel auf Bali]. *Geographische Zeitschrift* 87 (1), 29–45.

Watson, J. (1997) *Golden Arches East: McDonald's in East Asia*. Stanford, CA: Stanford University Press.

Weedon, G. (2012) 'Glocal boys': Exploring experiences of acculturation amongst migrant youth footballers in premier league academies. *International Review for the Sociology of Sport* 47 (2), 200–216.

Wherry, F.F. (2006) The social sources of authenticity in global handicraft markets – evidence from Northern Thailand. *Journal of Consumer Culture* 6 (1), 5–32.

Wieckowaski, M. (2010) Tourism development in the borderlands of Poland. *Geographia Polonica* 83 (2), 67–81.

Wilkins, W. (2011) Souvenirs – What and why we buy. *Journal of Travel Research* 50 (3), 239–247.

Williams, C.C. (2004) The myth of marketization: An evaluation of the persistence of non-market activities in advanced economies. *International Sociology* 19, 437–449.

Williams, C.C. and Nadin, S. (2010) Entrepreneurship and the informal economy: An overview. *Journal of Developmental Entrepreneurship* 15 (4), 361–378.

Williams, C.C., Nadin, S. and Barbour, A. (2010) *Making the Transition from Informal to Formal Enterprise: Barriers and Policy Solutions*. London: ISBE RAKE Fund.

Yang, L. (2011) Ethnic tourism and cultural representation. *Annals of Tourism Research* 38(2), 561–585.

Young, R.J.C. (2001) Preface. Sartre: The 'African Philosopher'. *Colonialism and Neocolonialism: Jean-Paul Sartre* (pp. vii–xxiv). London & New York: Routledge.

Yu, H. and Littrell, M.A. (2005) Tourist's shopping orientations for handcrafts: What are key influences? *Journal of Travel and Tourism Marketing* 18 (4), 1–18.

Zapata, M.J., Hall, C.M., Lindo, P. and Vanderschaeghe, M. (2011) Can community-based tourism contribute to development and poverty alleviation? Lessons from Nicaragua. *Current Issues in Tourism* 14 (8), 725–749.

Part 1

Theorising Experience and Behaviour

Part I

Forming Experience and Behaviour

2 With the Passing of Time: The Changing Meaning of Souvenirs

Noga Collins-Kreiner and Yael Zins

This chapter focuses on souvenirs and their changing significance through time, space and meaning. Today, souvenirs are closely related to a large number of cultural, social and economic phenomena, such as consumption and globalisation (Goss, 2004); identity, culture and materiality (Morgan & Pritchard, 2005); and shopping (Timothy, 2005).

The study documented here investigates time-dependent changes in tourists' attitudes toward their souvenirs, and examines the process by which objects change meanings in the eyes of their beholders. We begin with a review of the literature on souvenirs and continue with a presentation of methodology. We then present our findings, contextualised in a discussion of the characteristics and attitudes of the tourists under discussion, and outline the process by which objects become souvenirs. We conclude with a reassessment of the prevalent scholarly terminology on the subject.

The Meaning of Souvenirs

The subject of souvenirs has been explored extensively over the last decade in a number of fields, among them art, history, cultural studies and anthropology. Tourism researchers have also paid ample attention to souvenirs, as the phenomenon provides us with new insights about tourists, host populations and relationships between the two. Empirical studies exploring souvenirs have focused on the meaning of souvenirs (Gordon, 1986; Hitchcock, 2000; Shenhav-Keller, 1993), souvenir purchasers (Anderson & Littrell, 1996; Littrell *et al.*, 1994); issues of authenticity (Asplet & Cooper,

2000; Blundell, 1993; Littrell *et al.*, 1993); purchase intentions (Kim & Littrell, 1999, 2001); and travel motivations (Swanson & Horridge, 2006).

Many tourism and souvenir researchers have built upon Littrell's (1990) early benchmark study, which focuses primarily on objects and their meaning. Today, however, scholars argue that people in post-modern society are surrounded by material objects that can provide them with pleasure, security and refuge (Morgan & Pritchard, 2005; Wallendorf & Arnould, 1988). Objects convey special meaning to their owners and may even be sacred to them. Indeed, people's personal and social identities are linked to objects and objects acquired in the context of tourism or during tourist experiences are no exception (Haldrup & Larsen, 2006; Morgan & Pritchard, 2005).

The scholarship distinguishes between souvenirs and mementos. Souvenirs are commercial objects usually purchased during travel that remind us of past experiences and places visited. Mementos, in contrast, are not necessarily acquired during travel, and include non-commercial articles, or articles that were not produced or marketed to serve as souvenirs. Souvenirs may have individual and universal meanings, while mementos have only individual significance. Because souvenirs are commercial, many people deny the importance of souvenirs in their lives and instead attribute greater meaning to the mementos they have collected (Gordon, 1986). Stewart (1993) makes a similar distinction between external and internal appearances (souvenirs and mementos). The former are representative, mass produced and purchasable. The latter, in contrast, offer owners a mapping of their personal past, cannot be acquired as commercial products and are usually related in some way to a major rite of passage. This study treats souvenirs and mementos as synonymous, as they fulfill the same function in tourism.

Although the subject of souvenirs has been explored extensively in the last decade, the literature addresses neither time-dependant changes in souvenir meaning nor souvenir location within the homes of tourists, and in this way the present study makes a contribution to scholarship in the field. This study aims to fill a gap in the research by exploring the material importance of souvenirs, and owners' attitudes toward objects in the tourist context.

Methodology

The major objective of this study is to examine if and how the importance, value, quantity and location of souvenirs changes over time. Another objective is to consider whether the concept of souvenirs must be limited only to objects that are regarded as such by the literature or whether any item can become a souvenir.

The study employs quantitative and qualitative research methods in the form of surveys and in-depth interviews and observations. A diverse sampling of 211 Israeli tourists completed online surveys while traveling abroad. In order to participate, respondents had to meet two conditions: they had to have traveled abroad at least twice and at least three years had to have passed since their first trip. These requirements were meant to ensure that all participants could answer questions regarding time-dependent changes.

Surveys were completed online using a website designated for this purpose, and were distributed through the online forums of Israel's largest travel portal. This format obligated the respondents to answer all the survey questions. However, using an online format for completing the survey also had two major drawbacks. First, it was extremely difficult to determine the response rate, as it was impossible to determine the number of qualified potential respondents who entered the website and saw the referral to the survey. The second drawback was the method's inability to control respondents' age, which resulted in an uneven respondent age distribution that did not match that of the general population.

The survey consisted of four sections. The first section focused on souvenir acquisition patterns and included questions about acquisition habits and types of souvenirs acquired. The second section focused on respondents' travel histories and habits and included questions about the number of trips taken, travel frequency and style, destinations, travel motives and main interests during travel. The third section examined changes in souvenir acquisition patterns and attitudes towards souvenirs. The fourth and final section contained questions about respondents' socio-demographic characteristics.

The socio-demographic characteristics of the 211 participants were: male (42%) and female (58%); married (55%) and unmarried (45%); those with an academic education (more than 12 years of study resulting in an academic degree) (80%) and those without (20%); and those who considered themselves to be above the average income in economic status (50%) and those who considered themselves to be average or below average (50%). In terms of age, respondents fell into one of the following four groups: 18–25 years of age (18%); 26–35 years of age (51%); 36–50 years of age (20%); and 51 years of age and older (11%), the group that appeared to possess the most free income.

In addition to the surveys, 12 semi-structured, in-depth interviews were conducted with tourists in their homes. Interviewees from different socio-demographic backgrounds and from a wide range of ages (spanning eight decades) were chosen randomly. The diversity of the sampling provided us with more data to understand the phenomena. Of the interviewees, two-thirds (eight) were women and one-third (four) were men, while two-thirds were married and one-third were unmarried. A total of 50% of those interviewed had an academic education, while the other 50% did not. Findings were analysed qualitatively using the content analysis method.

The findings discussed in this chapter refer to the integrated findings of both questionnaires and interviews. We found it important to present the integrative work of the study regarding each subject, and not to separate our findings for each. For the purpose of clarification, however, interview-based data are followed by the interview number, while questionnaire-based data are not.

Findings

The changing meaning

Responses to the questionnaire revealed that 40% of the participants acquired objects that, although not intended to do so, nonetheless reminded them of their trip (this was one of the questions they were asked). We of course did not offer an explicit definition for the term souvenir, as one aim of this paper was to assess participants' subjective perception of the phenomenon.

These objects can be divided into two main categories: touristic articles and non-touristic articles. The first category included articles that are usually thought of as 'souvenirs' and mementos collected by tourists. The difference between these objects and ordinary souvenirs is the intention ascribed to them. While souvenirs and mementos are brought home to serve as a reminder of the journey, these objects are acquired for different reasons and only start functioning as souvenirs in retrospect. Examples of articles of this kind include: a location-specific key chain (which was broken but nonetheless retained as a souvenir); authentic houseware from France; local house decorations; local games; musical instruments; compact discs of local music; a mint candy wrapper from South Africa; a map of Namibia; a journal; local currency; lists made throughout the trip; entry tickets to sites; receipts from hotels and museums; maps and brochures; and other material objects that served a practical need while traveling but since then have acquired additional significance.

One respondent spoke of *'exotic fruit that dried up, which I kept,'* while another recalled *'a bookmark I got from a stranger I met on a ship.'* A third respondent kept *'An address of a man I had an interesting conversation with, who told me that "if you find someone that wants to invest in this country [Tanzania], this is the address." It reminds me of the villages and the interesting trip to get there,'* he explained.

The second category included articles of a non-touristic nature that started reminding owners of their trip only upon their return home or after a significant period of time had passed. These objects can be divided into two sub-categories. The first was objects acquired before departure that served the tourist throughout the trip, such as clothing and travel equipment, and objects purchased during the journey to be used along the way, such as: clothing, backpacks, dishes, scissors, toothpaste, laundry detergent, etc.

One respondent described such articles as follows in his interview: *'Simple objects I needed during the trip and that I bought along the way, such as good deodorant in Croatia, socks in Camden Town in London, good undershirts in Melbourne. Things I thought were meaningless and that I bought out of need, but which later I realised reminded me [of my trip] every time I wore them'.*

The second sub-category consisted of objects purchased throughout the journey to bring home, for example: *'Shoes and socks from New York that were purchased for practical use and became souvenirs'; 'Clothes I bought during my trip and didn't get rid of because they reminded me of a place';* jewelry and accessories (sunglasses and a watch); houseware; cosmetics; books; bags; and compact discs. Other objects mentioned included: writing and painting instruments; a purse; art books from museums; a camera purchased because of its low price; bags from shops; and bottles of wine *('empty wine bottles that I keep – I drank the wine').*

The changing importance

Most participants (60% of all participants) did not report a change in the importance of their souvenirs, while 40% did. Almost a third (29%) mentioned that the importance of their souvenirs diminished, while 11% said it increased (see Table 2.1). The primary reason for the diminished importance of souvenirs cited by respondents was an increased importance of photographs and mementos. *'I am more interested in souvenirs I get from people I meet along the way or things that I collect.'*

Other reasons given for diminished importance included the assertion that they were no longer required for remembering the journey (*'I understood that the memory is mostly in the mind and heart and not in the objects.'*); a realisation that the souvenirs themselves were actually unimportant (*'I discovered that when I returned home, the things that seemed important when I was there just sit around and collect dust.'*); a decline in sentimental value or meaning attached to the objects and decreased possessiveness (*'... because there is no real or sentimental value to it...';* *'Over the years, I feel less of a need to bring beautiful things I see into my possession.'*); and a decrease in their need to use purchases to increase social status (*'...maturing, buying only what you really need or want and not just to say "I have it"'*).

Table 2.1 Changes in souvenir location and importance

Variable	Responses	Frequency	%
Stable location of souvenirs over time	No	72	34%
	Yes	139	66%
Importance of souvenirs	Decreased	62	29%
	No change	127	60%
	Increased	22	11%

Increased tourist experience is a prevalent explanation for the diminishing importance of souvenirs. Having many souvenirs from past trips decreases the value of each souvenir, and confidence in future travel lessens tourists' need for souvenirs. One participant explained that *'over the years, going abroad ceased to be a unique experience that may never repeat itself or that I might not be able to afford again ... so there is no need for a souvenir of a "once in a lifetime experience".'* Another conjectured that perhaps this stemmed from *'...a feeling that the world has become smaller and you can return to places more easily, or see them on TV or on the internet.'* Many participants ascribed the change in their attitude to increased maturity and the passage of time. One explained that *'... they lack the magic of the moment when you saw them for the first time'*, while another said that *'the enthusiasm decreases with time.'*

Another reason for the diminishing importance of souvenirs was the increasing difficulty of finding items that are unique. Today, most things can be purchased in Israel: *'It is no longer as novel to go abroad as it used to be. Everything is more accessible, and even here you can buy all sorts of souvenirs from all over the world. It is not special anymore.'*

The changing location

Most study participants (66%) indicated that their souvenirs have remained in the same place over the years (see Table 2.1), while one-third (34%) indicated a change in their location within their homes. The main tendency has been to move souvenirs to less central locations within the home as time passes, eventually storing them in closets and boxes and in some cases throwing them away. A rare change in souvenir location noted by respondents was their transfer to a more prominent location in order to make better use of them.

It is difficult to assess all the factors that played a role in causing souvenirs to remain in their locations. Some may argue that this was the result of a decrease in a souvenir's importance, which caused it to fade into the background and to become such a regular part of the everyday scenery that it was never even moved. However, it can also be argued that such souvenirs remain important and that even though they become less significant to the tourist, the fact that they retain their place in the home nonetheless means something. In other words, they are not moved because their owners still want them in sight.

It is complicated to evaluate all the factors that led to changes in souvenir location. Some were the result of changes in the tourists' tastes, as souvenirs simply ceased to be aesthetically pleasing and tourists no longer wanted to display them in a central place in their homes, and in some cases sought to discard them. Occasionally, souvenirs became less important after tourists returned from subsequent trips. It should be noted, however, that the change in location of a souvenir does not necessarily indicate a change in the meaning it holds.

Location changes are occasionally required by reality, for instance after moving to a new residence or remodeling a home. In the case of remodeling, location change can be incidental, based solely on a desire for change and not necessarily on a change in attitude toward the souvenirs themselves. One interviewee explained that he regularly relocates different objects within his home 'as part of [his] practice of periodic redesigning.'

A few participants indicated that they left their older souvenirs in their parents' homes after moving out. This can be explained by a lack of space in the new home or by changed meaning. It may also represent a process of sorting, but instead of discarding the souvenirs they stored them somewhere out of the way. One interviewee explained that when she moved out of her parents' house she took most of her souvenirs with her but left behind those that no longer suited her taste or her new home. In her parents' home, the souvenirs were in no one's way, and some are still used to decorate her 'old bedroom'.

Another explanation for leaving souvenirs behind is spousal disapproval upon moving in together. Another interesting reason for change in souvenir location was noted by an interviewee, who explained that she sometimes moves a souvenir when she feels it is not in the 'right' place and that it would be better suited to a different location. In some cases, she brings souvenirs into her home gradually, initially placing them close to the entrance and subsequently moving them to more personal or intimate locations (such as her bedroom).

The changing quantity

In most cases, the number of souvenirs that participants acquired also changed over time. Thirty-eight percent of participants reported a decline in the number of souvenirs purchased; 49% reported no change; and only 13% reported an increase. In other words, just about half of the participants reported a change in the number of souvenirs acquired, with most buying fewer souvenirs as time passes. This trend was detected among the interviewees as well.

According to our findings, the main reason for change in the number of souvenirs purchased was the diminishing importance of souvenirs. In many cases, souvenirs that appeared meaningful during trips turned out to have little or no meaning once tourists returned home and time passed (see the discussion on this point below). In fact, it was often subsequent tourist experiences that reduced tourist enthusiasm for souvenir acquisition. One reason for this was the fact that many tourists returned to the same destination a number of times. Another was the fact that they already possessed similar souvenirs from other destinations. In the words of one respondent: *'If I go to Africa several times, how many animals [sculptures] do I need? Enough....'*

Tourists also appeared to become more particular on subsequent trips, seeking souvenirs that were more unique than those they had purchased in the past. As a result of all these factors, with time and, through calculated acquisition, respondents purchased a smaller number of souvenirs that were of higher quality on the whole.

Discussion

The findings show that objects gain and lose meaning after a journey is completed. This phenomenon is consistent with a broader, widespread decrease in the sentimental value of objects, and also appears to stem from repeated trips abroad and the transformation of international travel into a common event. With increased physical and virtual accessibility to the 'other,' travel for some has become less meaningful, and the significance attributed to souvenirs has decreased. For others, the trip and the memories are still extremely significant, but there is no longer the need for tangible souvenirs.

The study reveals that sentimental value can be assigned to seemingly meaningless articles, and vice-versa. That is to say, objects acquired explicitly as souvenirs often lose their sentimental value over time, while practical articles brought home from a journey often acquire sentimental meaning in retrospect. A total of 40% of study participants possessed such objects, demonstrating the importance of the phenomenon. The spectrum of objects that can become souvenirs is broad, and includes touristic articles, mementos that are not acquired to serve as a reminder of a trip but that begin fulfilling such a function at a later time, and articles that are not touristic in nature and that are acquired before, during or after a trip.

This study also reveals that after multiple journeys abroad, tourists acquired fewer souvenirs and ascribed less meaning to the ones they acquired during previous trips. It also shows that changes in souvenir location in the tourists' homes usually reflect changes in the meanings they ascribe to them. According to our findings, most participants do not change the location of souvenirs in their home, but the majority of those who do, move them to a less central place. In most cases, this reflects a decrease in the souvenir's importance in the eyes of the tourist. Some participants indicated that a souvenir's location does not necessarily reflect its significance. Others noted that as time passes, they feel that the memory has been engraved in their minds and that tangible souvenirs hold less significance. Still, for most participants, souvenir location within the homes indicated souvenir significance to some degree.

Time affects the meanings attributed to souvenirs, and these meanings, in turn, influence souvenir location within the home. The literature addresses neither time-dependent changes in souvenir meaning nor souvenir location within tourists' homes, and in this way the present study makes an important contribution to the scholarship.

With regard to the concept of micro-geography of souvenirs in the home, this study points to a solid pattern of movement from center to periphery. The souvenir enters the home together with the tourist and is positioned in a central place. With the passage of time, depending on different criteria – such as the age of the tourist, frequency of travel and souvenir importance – it is moved to the periphery of the house. After that, it is again moved and eventually disappears from view, suggesting that 'out of sight' is truly 'out of mind.'

This study found that souvenirs are dynamic and that they are often formed and transformed over time. Indeed, it is simply impossible to predict which objects will become souvenirs, as meaning is assigned in retrospect. The importance of this study is amplified by the lack of literature on the topics it addresses, particularly changes in souvenir location and souvenir meaning. Our findings, however, are consistent with the meager literature that does exist on the subject. Littrell *et al.* (1993) refers to situations in which objects can become souvenirs, based on the circumstances in which they are acquired.

We go one step further by arguing that every object has the potential for becoming a souvenir, based not only on the circumstances in which it is acquired, but also on a variety of other factors, such as the circumstances in which they were used. Changes in object meaning and the fact that, for many people, any journey-related object is a potential souvenir, indicates a need to expand the concept 'souvenir' as it is employed in the literature today. We posit that the term can be applied to every object that reminds us of a tourist journey, even if it was not originally acquired during travel and was not intended to serve this purpose from the outset.

This paper points to a fading distinction between our traditional conception of souvenirs and regular objects. The time has come for the contemporary usage of the term 'souvenir,' denoting a specific object for tourists, to expand in order to allow broader interpretations. Today souvenirs cannot be differentiated from other objects. The current approximation between objects and souvenirs is not only a consequence of a theoretical recognition of the homology between them, but also part of the tendency of contemporary, post-modern tourists to confuse the two during their trips more than in the past.

In addition to momentary experiences, tourism provides tourists with memories that are often treasured long after their return home. For this reason people need souvenirs – tangible symbols of their travels. This need has been intensified by the current material consumer culture. We argue that the term 'souvenir' must be rebranded to include not only objects that belong to the souvenir typologies existing today, but also more ordinary objects that begin functioning as souvenirs later, after the tourist's return home. Indeed, many tourists regard this kind of souvenir as the most significant kind.

Summary

The overall importance of objects has already been identified in the tourism and souvenir literature. This article contributes to current knowledge by analysing how souvenir meaning and location change over time. It aims to fill a gap in the research by exploring the material importance of souvenirs and changes in tourists' attitudes toward objects over time.

The study reviewed in this chapter also reveals a need to reassess and revise the academic definition of the term 'souvenir' and suggests that the term should be defined in a broader manner to include any object that serves as a reminder of a journey, even if that object was not intended to fulfil this function. Indeed, the concepts discussed here, such as the possibility of co-existence of a multiplicity of definitions of souvenirs rather than the victory of only one, are consistent with today's trends in research.

References

Anderson, L.F. and Littrell M.A. (1996) Group profiles of women as tourists and purchases of souvenirs. *Family and Consumer Sciences Research Journal* 25 (1), 28–57.

Asplet, M. and Cooper M. (2000) Cultural designs in New Zealand souvenir clothing: The question of authenticity. *Tourism Management* 21, 307–312.

Blundell, V (1993) Aboriginal empowerment and souvenir trade in Canada. *Annals of Tourism Research* 20, 64–87.

Gordon, B. (1986) The souvenir: Messenger of the extraordinary. *Journal of Popular Culture* 20 (3), 135–146.

Goss, J. (2004) The souvenir: Conceptualizing the object(s) of tourist consumption. In A.A. Lew, M.C. Hall and A.M. Williams (eds) *A Companion to Tourism* (pp. 327–336). Malden, MA; Oxford, UK; Carlton, AU: Blackwell Publishing.

Haldrup, M. and Larsen, J. (2006) Material cultures of tourism. *Leisure Studies* 25 (3), 275–289.

Hitchcock, M. (2000) Introduction. In M. Hitchcock and K. Teague (eds) *Souvenirs: The Material Culture Of Tourism* (pp. 1–15). Aldershot: Ashgate.

Kim, S. and Littrell M.A. (1999) Predicting souvenir purchase intentions. *Journal of Travel Research* 38 (2), 153–162.

Kim, S. and Littrell, M.A. (2001) Souvenir buying intention for self versus others. *Annals of Tourism Research* 28 (3), 638–657.

Littrell, M.A. (1990) Symbolic significance of textile crafts for tourists. *Annals of Tourism Research* 17 (2), 228–245.

Littrell, M.A., Anderson, L.F. and Brown, P.J. (1993) What makes a craft souvenir authentic? *Annals of Tourism Research* 20 (1), 197–215.

Littrell, M.A., Baizerman, S., Kean, R., Gahring, S., Niemeyer, S., Reilly, R. and Stout, J-A. (1994) Souvenirs and tourism styles. *Journal of Travel Research* 33 (1), 3–11.

Morgan, N. and Pritchard, A. (2005) On souvenirs and metonymy, narratives of memory, metaphor and materiality. *Tourist Studies* 5 (1), 29–53.

Shenhav-Keller S. (1993) The Israeli souvenir – its text and context. *Annals of Tourism Research* 20 (1), 182–196.

Stewart, S. (1993) *On Longing: Of the Miniature, the Gigantic, the Souvenir, the Collection.* Durham and London: Duke University Press.

Swanson K.K. and Horridge P.E. (2006) Travel motivations as souvenir purchase indicators. *Tourism Management* 27 (4), 671–683.

Timothy, D.J. (2005) *Shopping Tourism, Retailing and Leisure.* Clevedon: Channel View Publications.

Wallendorf, M. and Arnould, E.J. (1988) 'My favorite things': A cross-cultural inquiry into object attachment, possessiveness and social linkage. *Journal of Consumer Research* 14 (4), 531–547.

3   Souvenirs and Self-identity

Hugh Wilkins

When we think about souvenirs we normally do so in the context of the event that led to their creation. Souvenirs, or mementos, are items that remind us of a significant event or time in our lives and include both paid for items, and collected or free items. They are normally tangible, although with the development of technology the potential for intangible souvenirs, such as digital images and other recordings, must be recognised. The purchase of mementos and souvenirs is a well-established behaviour associated with many activities, including vacations and other leisure activities. Few people will undergo a vacation without acquiring some form of evidence to tangibilise the experiences gained (Gordon, 1986). The type of evidence will vary from person to person and from experience to experience, and from a T-shirt to a tapestry or a shell to a snow-scene. Gordon (1986) goes beyond this by stating that 'almost everyone is, in one way or another, a collector of souvenirs. People like to be reminded of special moments and events'.

Tourism has been described as a 'sacred journey' (Graburn, 1979: in Gordon, 1986) with there being a need for people to bring back mementos and souvenirs of the sacred, extraordinary time or space, not only to aid recollection of the experience, but also to prove it (Gordon, 1986). The gathering of souvenirs is, therefore, a means of making tangible an experience, either for consumption by others, or as means of prolonging the experience for one's own consumption. As MacCannell (1976: in Gordon, 1986) stated 'tourists return home carrying souvenirs and talking of their experiences, spreading, wherever they go, a vicarious experience'.

Although not all souvenirs are purchased items, the majority are commercial products and shopping is a major tourist activity (Fairhurst et al., 2007; Kim & Littrell, 2001). Souvenir purchases form a significant component of the shopping expenditure (Lehto et al., 2004; Littrell et al., 1994) and previous research has suggested expenditure on shopping comprises almost a third of the total travel spend (Fairhurst et al., 2007; Littrell, et al., 1994). Much of the research into souvenir purchase behaviour has focused on the characteristics of the purchaser, e.g. Anderson and Littrell (1996), Kim and Littrell (2001), or the authenticity of the purchase (Littrell *et al.*, 1993), rather than the motivational aspects for the purchase decision. This research is an initial exploration of these motivational aspects.

The concept of congruence (Dolich, 1969; Martin & Bellizzi, 1982; Sirgy, 1985) suggests that consumers tend to buy products that reflect actual or aspirational self-perception, partly at least because the products communicate to others details about lifestyles. Leisure activities, being freely chosen, have greater expression of self-identity and thus may be seen as more important in defining the self (Haggard & Williams, 1992). They are especially high in symbolic content, with them providing an opportunity for self-expression and social positioning (Dimanche & Samdahl, 1994). Leisure activities are especially likely to be affected by self-congruity factors, but the literature would suggest that the correlation is more attuned to the congruity between the ideal or ideal social self and the activity, as the activity itself is used as a means of self-expression. Given the established role of congruence it would be reasonable to assume that congruence also applies to the purchase of souvenirs, not because souvenirs themselves are reflective of the consumer, but because the souvenir is reflective of the vacations or other leisure activities we have undertaken.

This importance is extended because previous research has also shown consumers use vacations as a form of materialism (Wilkins, 1998) and that, combined with the role of the souvenir as evidence (Wilkins, 2011), supports the argument for the extended role of souvenirs by Shenhav-Keller (1993) who identifies a souvenir as being symbolic and having implicit meanings that 'can be read as a text' (Shenhav-Keller, 1993: 183). Consumers may buy, own, use and display certain products and services, including souvenirs, to enhance their sense of self and to present an image of who they are (Eastman *et al.*, 1999). On the basis of this external role of the souvenir it is important to evaluate the relationships between the souvenir, self-image and product congruence, all of which are related to the travel experience.

The previous research has focused on several aspects of purchase behaviour associated with souvenirs and shopping expenditure by tourists, but there has been little focus on the motivations for souvenir purchase. This research extends previous work by focusing on both the motivations for purchase and the relationship to self-identity.

Method

A mixed method approach was adopted with initial qualitative data collection being supported by a quantitative stage addressed through a self-completion questionnaire. Items for the survey instrument were generated both inductively and deductively.

Three focus groups were held, each lasting between 60 to 90 minutes and being recorded. The recordings were transcribed, and analysed using Nud*ist, a qualitative analysis software program. The data from the focus groups was used to inform the development of items for inclusion in a self-completion survey. These items were combined with items generated from the literature.

The survey instrument was subject to a pilot study to purify the dimensions contained in the survey instrument. The questionnaire contained four sections related to vacations and vacation activities; souvenir attributes and purchase motivations; personal attitudes; and demographics. The first section of the survey was designed to obtain information about respondents' vacation preferences, particularly in relation to vacation styles, selection and activities. The second section collected information regarding souvenirs, such as the types of souvenirs purchased, attributes and purchase motivation. The third section comprised questions relating to respondents' personal values in relation to vacation styles, and included questions on preferred holiday styles and activities. The fourth section comprised demographic questions, such as respondents' gender, age, country of residence and income.

A web-based data collection approach was adopted. The survey was converted into a web-based format using SPSS Data Entry Web Builder 3.0, and a website was designed for the project. The website consisted of four components, introductory letter, sponsor information and links, link to survey instrument and link to prize draw information.

To collect responses, a 'snowball' technique was applied. 'Snowball' or 'network' sampling occurs when the researcher identifies one member of the population, and then other members are identified by this member, and so on (Jennings, 2001). In initiating this technique, an email was sent to all students at Griffith University in Queensland, Australia and a range of business contacts asking them to complete the online survey and to forward the email onto family, friends and colleagues. The email gave a brief description of the research project and provided a hotlink to the web address. In total 3231 completed and usable survey responses were received.

The items relating to souvenirs comprised two scales in the survey instrument. The first scale titled 'Souvenir Attributes' contained items directly related to the souvenir characteristics, for example 'the design is appealing' or 'the workmanship or techniques are of high quality'. This scale comprised 12 items. The second scale addressed purchase motivations and was split into two sections. The scale included 27 items. The first section contained items derived solely from the focus groups and included items such as 'souvenirs are a reminder of how special my travel experiences were' and

'I like to buy souvenirs that I can be used as Christmas or birthday presents'. The second section was based on scales derived from the literature and included items such as 'I like to buy souvenirs that identify where I've been' and 'I buy souvenirs from local artists'.

As there was some similarity between individual items in the different scales, all 39 items were placed into a single factor analysis to identify the naturally occurring dimensions. This approach is recommended in the literature as a means of identifying actual, rather than perceived, factor groupings (Rosen & Surprenant, 1998).

Findings

The 39 items were subject to an exploratory principal components analysis using an oblique rotation. An oblique rotation was used as it is a recommended approach when correlation between components is anticipated (Hair et al., 1998). As the sample size exceeded 350, a factor loading of 0.40 was deemed acceptable (Hair et al., 1998). Each scale was analysed and items that were complex, that is, loaded strongly onto more than one component, or showed a component loading below 0.40, were removed.

A three-component solution was identified. The Cronbach alpha for the resultant scales was used to measure the scale reliability. The scale structures were then subjected to confirmatory factor analysis. As recommended by Bagozzi and Heatherton (1994), a partial disaggregation was used, with each dimension represented as a separate latent variable measured by composites of subscales (Bagozzi & Heatherton, 1994; Williams & O'Boyle Jr, 2008). The use of a partial disaggregation model provides a more detailed representation of constructs and supports the evaluation of discriminant validity (Churchill, 1979).

Owing to the sensitivity of fit indices to sample size, a random sample of 300 respondents was selected from the data to ascertain level of fit (Marsh et al., 1988). The confirmatory factor analysis fit indices confirmed the structure, with all results within the recommended ranges. A second random sample of 300 respondents was then selected from the remaining data to retest the results. The structure of the dimension was supported by the retest sample. Both convergent and discriminant validities were established. Convergent validity was demonstrated with inter-dimension correlations all significant ($p < 0.01$) (Bagozzi et al., 1991). Discriminant validity was also established with the Average Variance Extracted (AVE) for each dimension being greater than the square of the inter-correlations (Fornell & Larcker, 1981). The fit indices for the confirmatory factor analysis are provided in

Table 3.1. The table provides the results for the overall dimension and the individual constructs.

It should be noted that while memory is inwardly focused, that is on recreating the experience for the participants, the other two motivations are externally focused, that is they are for consumption by other people.

In addition to the confirmatory factor analysis the data were also analysed using hierarchical multiple regression to identify the effect on the souvenirs purchased. This analysis enables the inclusion of direct and moderating effects as sequential components, thus enabling the identification of the additional explanation provided by the moderating variables (Aguinis, 1995; Francis, 2004; Hair et al., 1998; Pedhazur, 1997; Tabachnick & Fidell, 2001). The use of hierarchical multiple regression or incremental partitioning of variance (Pedhazur, 1997) is a recommended approach in moderated multiple regression, in which the impact of a moderating variable on the relationship between two variables is measured (Aguinis, 1995).

As can be seen in Table 3.2, the predictors of evidence, memory and gift as purchase motivations gain a R^2 of 0.292 for the souvenir purchase choice. It should be noted that the role of evidence is the principal influence of the souvenir selected. The second most important factor was the role of gift, and memory was not significant for males.

Table 3.1 Confirmatory factor analysis fit indices

Dimension				Norm				
	χ^2	DF	P	χ^2	AGFI	CFI	RMSEA	SRMR
Gift	4.54	2	0.103	2.27	0.963	0.993	0.065	0.018
Memory	4.78	2	0.092	2.39	0.962	0.996	0.068	0.014
Evidence	10.27	5	0.068	2.05	0.960	0.988	0.059	0.026

(After Wilkins, 2011)

Table 3.2 Predictors of evidence, memory and gift as purchase motivations

Model	R	R square	R square change	F change	Sig. F change
1	0.490[a]	0.240	0.240	1020.046	0.000
2	0.532[b]	0.283	0.043	191.756	0.000
3	0.541[c]	0.292	0.009	43.104	0.000

[a] Predictors: (constant), evidence
[b] Predictors: (constant), evidence, gift
[c] Predictors: (constant), evidence, gift, memory
(After Wilkins, 2011)

The first factor contained six items and was named *Evidence* as the items were related to the showing of the souvenirs. Examples of the items include *I like souvenirs you can talk about with others* and *I like to put my souvenirs on display to show the places I have visited*. These items reflected data collected in the focus groups where people were explicit about the role of the souvenir as evidence of their travels with, for example, one respondent stating 'I do buy a lot of souvenirs ... that I will display' and another, in reference to a piece of jewellery said 'and people say "oh, that's nice" and you have the opportunity of talking about your holiday'. Although the intention of these souvenirs are as evidence, it was also stated that people did not want to use them for bragging with another respondent commenting 'But it's nice when someone says "Where did you get such and such – it's lovely". Yes, if they say it first then it can be talked about'. The Cronbach alpha for the scale was 0.841.

The second factor also contained six items and was named *Memory* as the majority of the items in the component reflected the role of bringing back the travel experience. Examples of these items include *the souvenirs I buy bring connection to my trip* and *Souvenirs are a reminder of how special my travel experiences were*. These items, derived from the focus groups, reflected confirmation of a strong theme in the focus groups of the souvenirs bringing back memories of better, or more exciting times, with one respondent commenting 'they (the souvenirs) remind me of when I had a life'. The Cronbach alpha for this dimension was 0.909.

The final factor was titled *Gift* as it related to the sharing of the souvenirs with others. This component contained seven items and the Cronbach alpha for this dimension was 0.754. The items in this dimension related to the use of souvenirs as gifts with it containing items such as *I like to buy souvenirs that I can give to family or friends as gifts for special occasions*.

Composite variables were created for each of the identified factors. A correlation matrix was obtained of the components against the four elements of self-identity; actual self, ideal self, actual social self and ideal social self. These correlations are provided in Table 3.3.

Discussion

Although, with one exception, all aspects of the self have significant correlations with each composite factor ($p < 0.01$) it is important to note the differences. An examination of the *Evidence* factor shows relatively strong correlations with all aspects of the self, although the correlations are noticeably higher with the *Ideal* dimensions of the self. The correlation with *Ideal Self* (0.199) is strongest, with this implying the souvenirs we buy as evidence,

Table 3.3 Correlations between souvenir purchase motivations and self-identity

	Actual self	Ideal self	Actual social self	Ideal social self
Evidence	0.149**	0.199**	0.131**	0.176**
Memory	0.199**	0.164**	0.024	0.054**
Gift	0.132**	0.159**	0.109**	0.140**

**Correlation is significant at the 0.01 level (two-tailed)

which are the ones that are put on display, are souvenirs that reflect our ideal self-image, as opposed to the *Actual Self* (0.149). This pattern is also mirrored with the social self-dimension correlation with *Ideal Social Self* being stronger than the correlation with *Actual Social Self*. This implies a rational behaviour pattern, with consumers selecting evidential souvenirs to reflect ideal, rather than actual self-image.

An examination of the second factor, *Memory*, interestingly shows a very different pattern with the strongest correlation being with *Actual Self* (0.199), stronger than with *Ideal Self* (0.164). However the correlations with social self are very low and this also implies rational purchase behaviour of souvenirs with people buying souvenirs designed to be kept as a personal 'aide de memoire' being most reflective of the actual self, that is reflective of the real person.

The final factor, *Gift*, more closely follows the pattern seen for *Evidence* with the *Ideal Self* being stronger than the *Actual Self*. In a similar interpretation to *Evidence*, it would seem reasonable that consumers again select souvenirs as gifts that will impress the recipient and act as an aspirational indicator.

Conclusion

This research has explored the motivations associated with souvenir purchase and has identified three dimensions of motivations for purchase. The increased understanding of the souvenir purchase motivations will be of value to the tourism industry, as an increased understanding of the motivational dimensions will enable better segmentation and positioning of products. The importance of shopping as a component of vacation expenditure is high (Littrell *et al.*, 1994). The expenditure on souvenirs forms an important component of shopping expenditure and research that enables retailers to more effectively manage their product range and to communicate the attributes of this to potential customers is important in this competitive sector of the industry.

The evaluation of the relationship of these dimensions with self-identity provides new understanding of the aspects of the self-identity contributing to souvenir purchase motivation. Of particular relevance is the implication about how consumers differentiate in their souvenir purchases depending on the purpose of the souvenir, and this provides a new understanding of souvenir purchase motivations and behaviours. The increased understanding provided by the interrelationship between self-identity and souvenir purchase motivation will better enable retailers to promote and communicate product attributes.

References

Aguinis, H. (1995) Statistical power problems with moderated multiple regression in management research. *Journal of Management* 21 (6), 1141–1158.

Anderson, L. and Littrell, M.A. (1996) Group profiles of women as tourists and purchasers of souvenirs. *Family and Consumer Sciences Research Journal* 25 (1), 28–55.

Bagozzi, R.P. and Heatherton, T.F. (1994) A general approach to representing multifaceted personality constructs: Application to state self-esteem. *Structural Equation Modelling* 1 (1), 35–67.

Bagozzi, R.P., Yi, Y. and Phillips, L.W. (1991) Assessing construct validity in organizational research. *Administrative Science Quarterly* 36 (3), 421–458.

Churchill, G.A. Jr (1979) A paradigm for developing better measures of marketing constructs. *Journal of Marketing Research* 16 (1), 64–73.

Dimanche, F. and Samdahl, D. (1994) Leisure as symbolic consumption: A conceptualization and prospectus for future research. *Leisure Sciences* 16 (2), 119–129.

Dolich, I.J. (1969) Congruence relationships between self images and product brands. *Journal of Marketing Research (pre-1986)* 6 (000001), 80–80.

Eastman, J.K., Goldsmith, R.E. and Flynn, L.R. (1999) Status consumption in consumer behavior: Scale development and validation. *Journal of Marketing Theory and Practice* 7 (3), 41–52.

Fairhurst, A., Costello, C. and Holmes, A.F. (2007) An examination of shopping behavior of visitors to Tennessee according to tourist typologies. *Journal of Vacation Marketing* 13 (4), 311–320.

Fornell, C. and Larcker, D.F. (1981) Evaluating structural equation models with unobservable variables and measurement error. *Journal of Marketing Research* 18 (1), 39–50.

Francis, G. (2004) *Introduction to SPSS for Windows*. Frenchs Forest: Pearson Education Australia.

Gordon, B. (1986) The souvenir: Messenger of the extraordinary. *Journal of Popular Culture* 20 (3), 135–151.

Haggard, L.M. and Williams, D.R. (1992) Identity affirmation through leisure activities: Leisure symbols of the self. *Journal of Leisure Research* 24 (1), 1–18.

Hair, J.F. Jr, Anderson, R.E., Tatham, R.L. and Black, W.C. (1998) *Multivariate Data Analysis* (5th edn). Englewood Cliffs, New Jersey: Prentice Hall.

Jennings, G. (2001) *Tourism Research*. Milton: John Wiley & Sons.

Kim, S. and Littrell, M.A. (2001) Souvenir buying intentions for self versus others. *Annals of Tourism Research* 28 (3), 638–657.

Lehto, X.Y., Cai, L.A., O'Leary, J.T. and Huan, T-C. (2004) Tourist shopping preferences and expenditure behaviours: The case of the Taiwanese outbound market. *Journal of Vacation Marketing* 10 (4), 320–332.

Littrell, M.A., Anderson, L.F. and Brown, P. J. (1993) What makes a craft souvenir authentic? *Annals of Tourism Research* 20, 197–215.

Littrell, M.A., Baizerman, S., Kean, R., Gahring, S., Niemeyer, S., Reilly, R., *et al.* (1994) Souvenirs and tourism styles. *Journal of Travel Research* 33 (1), 3–11.

Marsh, H.W., Balla, J.R. and McDonald, R P. (1988) Goodness-of-fit indexes in confirmatory factor analysis: The effect of sample size. *Psychological Bulletin* 103 (3), 391–410.

Martin, W.S. and Bellizzi, J. (1982) An analysis of congruous relationships between self-images and product images. *Journal of the Academy of Marketing Science* 10 (4), 473–489.

Pedhazur, E.J. (1997) *Multiple Regression in Behavioral Research: Explanation and Prediction* (3rd edn). Forth Worth: Harcourt Brace.

Rosen, D.E. and Surprenant, C. (1998) Evaluating relationships: Are satisfaction and quality enough? *International Journal of Service Industry Management* 9 (2), 103–125.

Shenhav-Keller, S. (1993). The Israeli souvenir: Its text and context. *Annals of Tourism Research* 20, 182–196.

Sirgy, J.M. (1985) Using self-congruity and ideal congruity to predict purchase motivation. *Journal of Business Research* 13 (3), 195–206.

Tabachnick, B.G. and Fidell, L.S. (2001) *Using Multivariate Statistics* (4th edn). Boston: Allyn and Bacon.

Wilkins, H. (1998) A new perspective on materialism. *Tourism, Culture & Communication* 1 (2), 109–114.

Wilkins, H. (2011) Souvenirs: What and why we buy. *Journal of Travel Research* 50 (3), 239–247.

Williams, L.J. and O'Boyle, E.H. Jr (2008) Measurement models for linking latent variables and indicators: A review of human resource management research using parcels. *Human Resource Management Review* 18 (4), 233–242.

4 Souveniring Occupational Artefacts: The Chef's Uniform

Richard N.S. Robinson

When one thinks of souvenirs, the immediate image is of a tacky touristic trinket that somehow connects the travelled collector with their previous experiences. Memory of travels and evidence of having 'been somewhere' in particular, are strong motivators for souvenir acquisition (Wilkins, 2011). For the international traveller this could be construed as the 'global made local' – authenticating, reliving and recalling experiences using the souvenir as stimuli. Traveller experiences however, are produced, or created, by a collective global workforce, either directly or indirectly employed in tourism, estimated at 238 million (WTTC, 2011). One of the occupational groupings integral to producing the tourist experience is that of the chef – a calling enigmatically at once both marginal yet celebrated (Wood, 2000). From a traveller's perspective cookery, or food-related (tangible), souvenirs might include menus, various merchandise including celebrity chef cookbooks, branded aprons and kitchen utensils, even the odd illicitly acquired item of cutlery or other table accoutrements! One rarely considers however, the souveniring behaviours of the occupation itself, and how and why the collection of occupational artefacts relates to a local–global nexus.

'Iron Chef', is one of the iconic and cult cookery programmes of the late 20th and early 21st century (see Gallagher, 2004). Produced by Fuji Television, in Japan the game-show format pits renowned Oriental chefs against each other in a dramatic, busily camera-panned cook-off, sound-tracked by the baritone gasps of heartthrob host, Takeshi Kaga, and the effusive squeals of invited celebrity judging panellists. Conjuring up culinary creations that often usurps the dominant 'East meets West' fusion cuisine, the contestant chefs are predominantly nationals recruited from leading Japanese, Chinese and Korean hotels and restaurants. While the occasional Caucasian chef features, invariably one specialising in an Oriental cuisine, why do the chefs, while showcasing rich Asian gastronomic heritages, attire in the traditional Western (Francophile to be accurate) chefs uniform? Perhaps it is because the upmarket restaurant and hotel industry in Asia is structurally akin to its European counterparts? Or perhaps the show's producers are marketing 'Iron Chef' to an international audience? Either way, one questions the global symbolism of the chefs' uniform and its components. This chapter explores the souveniring behaviours of a sample of chefs, a sometimes marginalised occupation (Robinson, 2008) oftentimes working on the margins of the tourism industry, and this behaviour as a glocal phenomena. Specifically, this chapter focuses on what the uniform means to a sample of chefs who practice and live in Australia – thousands of kilometres away from the 'Iron Chef' studios. In particular, this investigation focuses on whether occupational artefacts are souvenired at the local level and whether they are referential at the global level.

Tourism, Food and Chefs

All tourists must eat. A conservatively estimated 25% of tourism expenditure is attributable to food products (Correia et al., 2008). Food consumption is key to creating memorable experiences in hospitality contexts (e.g. Lashley et al., 2003), but is also acknowledged to positively influence tourist experiences of a destination (Wolf, 2006) and in generating satisfactory travel experiences. As a cultural artefact, food provides a medium for the expression of local culture and connects tourists with a destination's landscape and unique way of life (Ottenbacher & Harrington, 2013). Therefore, food is an important destination attribute.

Many tourism destinations now promote themselves as centres of food and culture (Okumus et al., 2007) and destination restaurants are a feature of many of these (Sparkes et al., 2003). There is evidence to suggest 'foodies' plan travel itineraries around food experiences – those created principally by chefs. This phenomena itself suggests that a rich insight may be gained from investigating the souveniring of tangible food tourism artefacts, for example menus, and also intangible souvenirs, for example sensory experiences that may trigger the recollection of experiences in the future (Lupton, 1994). This chapter however, is not concerned with food as a cultural artefact, nor with how travellers souvenir various food-related artefacts, but rather with the artefacts of the occupation that by and large creates the food experiences of travellers, and for this purpose something must be understood of the nature of work of chefs.

Occupational culture and community

A range of contemporary media, either tacitly or overtly, provide portrayals of a chef's work and in so doing inform perceptions of occupational identity. Celebrity is a phenomenon that has firmly attached itself to the occupation of cookery (Wood, 2000). In contrast to the pervasive celebrity identities of 'Iron Chefs', Gordon Ramsay, Jamie Oliver (Gunders, 2008) and countless others, there are several identity theories, for example social worlds theory (Elkjaer & Huysman, 2008) and social identity theory (Tajfel, 1974; Turner, 1975), which suggest that occupational communities have sociological rather than organisational behaviour origins. Lee-Ross (2005: 3) defines occupational communities and describes their determinants as:

> work-based self-image where members see themselves in terms of their occupational role and where their self-image is based upon their occupational role. Self-image is valued by its holders and is shared by the rest of the occupational community; a work-based reference group composed of occupational community members who are socialized to share attitudes, viewpoints and values; work-based friends, interests and hobbies.

There are many features of a shared occupational community, or culture, of chefs, which both differentiates them from other occupations but are also used by the community to forge identity, inclusions and exclusions. This later point is a moot one for hospitality industry workers in general (Wood, 1992). Chefs (Palmer et al., 2010), have been traditionally celebrated as a marginal occupation. Themes such as violence in the kitchen (e.g. Murray-Gibbons & Gibbons, 2007), recruitment from the margins (Mullen, 1997) and deviant behaviour (Harris & Ogbanna, 2002; Robinson, 2008) resonates in the chef literature and accentuate the paradox between a localised marginalised profession and global celebrity.

Nonetheless, other common characteristics that shape, and are shaped by, the occupation are apparent in the literature. These include creativity (Cameron, 2001; Peterson & Birg, 1988), language (Fine, 1996), training (Cameron, 2001), knowledge of nutrition (Middleton, 2000), communication (Murphy & Smith, 2009), humour (Lynch, 2009), gender construction (White et al., 2005) and mobility (Robinson & Beesley, 2010). Yet no literature to date directly investigates the physical artefacts, as opposed to behavioural characteristics, which contribute to a strong, arguably global community, of chefs bound by a shared understanding of occupational artefacts.

Occupational artefacts

Sociological and behavioural literatures into occupational artefacts are broadly founded on the notion that objects are symbolic and convey membership and identity. Goffman (1959) contextualised the import of objects to the hospitality industry in his dramaturgical metaphor. The uniform is a highly visible occupational artefact and Wood (1966) suggests it is a phenomenon that has organisational, individual and systemic meaning. Within a hotel the roles prescribed to (key operational) departments is clearly delineated by uniform style. Individual 'actors' seek identity and membership according to their personal uniform, regardless of whether it is provided by the organisation or not, as can be the case (Rowley & Purcell, 2001). The meta-labels 'white collar' and 'blue collar' are clearly indicative of complex systemic structures borne of prescriptive as well as informal workforce attire, thus a uniform is a clear symbol of what is expected in various social contexts (Ashforth & Greiner, 1999).

The chef's uniform

The most visible occupational artefact of the chef is the uniform and there is a good deal of mythologising regarding its origins. The uniform's components (from top down) are the *toque blanche* (chef's hat), neckerchief, white double-breasted jacket with white or black buttons, a long white waisted apron with a *torchon*, or dish cloth, tucked into it, black and white checked trousers (traditionally in a tight hounds tooth pattern) and boots, with clogs being a classical style (see Figure 4.1). There are many practical reasons explaining the uniform's features. For example the *toque* (partially) prevents hair finding its way into preparations, the neckerchief mops up sweat, the double-breast of the jacket can be reversed to disguise the evidence of a chef's labour when ceremonial duties call and the tight trouser checks obscure stains. These less than romantic considerations are in contrast with stories relating the origins of the uniform.

Many stories revolve around the *toque*. One tale traces the origins of the *toque* to the paranoia of an ancient Assyrian king, who insisted on the court's cooks wearing a heavily pleated headdress, similar to that of the royals, so that infiltrators wishing to poison him would be easily identifiable. Another myth is that during antiquity a laurel leaf-garnished bonnet was presented to cooks before official feasts (The Culinary Institute of America, 2001). A common theme explaining the origin of the *toque* is a monastic heritage revolving around the Byzantine Empire, which straddled the East and West in various manifestations for a millennium. One legend suggests that persecuted philosophers and artists sought refuge in the Orthodox Church

Figure 4.1 The chefs' uniform

monasteries by becoming cooks and donned the robes and headdress of the monks to avoid detection, and over time the cooks chose white to distinguish themselves from the religious (Sockrider, 2005).

There is more tangible evidence, much in art and literary work, to suggest that in post-Renaissance Europe a variety of *toque* and uniform styles were adopted. Careme is said to have brought the *toque* into the professional restaurant kitchen, which housed many former royal chefs made redundant during the French Revolution. These former chefs of the courts retained their military-reminiscent jackets, which again are said to evolve stylistically from monastic robes (Sockrider, 2005). The other components of the chef's uniform became gradually standardised during the late Victorian period and through the 20th century, such that they are now globally instantly recognisable icons for the occupation of cookery.

The occupational artefacts of chefs, however, extend beyond apparel, or the uniform. Items of equipment, for example knives, steel (see Figure 4.1 again), tools, gadgets, scroll, toolkit and/or knife box come to mind. Equipment specific to an occupation also has meaning attached. Should the objects have been created, then they embody an occupationally unique knowledge referential beyond the individual or organisation in which they reside. Other artefacts serve as boundary markers (Carlile, 2002), for example the chef's tool box, which signifies 'a worker'. In this way, more than just technically functional, occupational artefacts are of social import (Bechky, 2003). Moreover, various kitchen management tools, such as photographic records, menus, recipes and rosters, also constitute legitimate occupational artefacts. Other functional documents like the resume and reference letters can be complemented by memorabilia, including awards, medals, certificates, props for buffets, displays, platters, dishes and the like. Despite this plethora of possible occupational souvenirs, the study presented below found the chefs uniform to be a most significant artefact, also associated with souveniring.

Data for this study is derived from a concentrated series of 52 semi-structured in-depth interviews with chefs and former chefs conducted in Brisbane, Australia, between 2005 and 2010. Each interview followed the same protocol, which consisted of four key questions.

(1)   What were the circumstances surrounding your entry into professional cookery?
(2)   What do you like/dislike about your work?
(3)   Why have you left previous positions/your occupation? And
(4)   What are your future goals?

Theoretically, the first question probed occupational selection and socialisation issues, the second question job and occupational satisfaction, the third question intentions and decisions to quit and the final question career planning and work/life goals. The researcher's emic status allowed for a rich rapport to develop, and interviews manifest as conversations and participant [pseudo]names are used in reporting the findings to highlight this relationship. This rapport was important since the study's underpinning construct was occupational culture and community. Indeed, clothing and equipment has been probed to unveil these in previous studies (e.g. Palmer et al., 2010). It was in this way that discussion of the artefacts of the occupation, their souveniring, relevance, importance and symbolic significance, emerged.

Many of the chefs interviewed had worked across the globe and related experiences as diverse as working as private chefs in royal palaces and for the exceedingly wealthy on private luxury yachts. Others worked in Michelin star and award-winning restaurants and in boutique gentlemen's clubs, and some had competed at prestigious international *salon culinaire* competitions. Many had practiced with a range of international hotel and resort chains, on luxury cruise liners and national institutional caterers. At the other end of the tourism and hospitality industry spectrum, the participants recounted working in local hotels, suburban cafes, franchised coffee shops, pizzerias, in sporting, community and striptease clubs and even desert truck stops. At time of interview, 37 of the participants were still practicing and two were employed as cookery trainers. Of the remainder, 12 had retired from professional cookery, although three of these retirees still worked in the hospitality or auxiliary industries. Finally, one chef was on maternity leave.

The sample was predominantly Australian (63%), but British, Irish, French, German, Hungarian, Belgian, South African, Singaporean, Ecuadorian and New Zealand-born chefs were represented. Males comprised 60% of the sample and females 40%. The age of participants, at time of interview, ranged from 19 to 70, and the median age group was 35–40. Collectively, the sample had over 800 years experience working as professional chefs and had practiced across Australasia, in the UK, Europe, the Americas, Asia and the Middle East. In summary, the participants, their countries of origin and their diverse industry experiences, juxtaposed with the context of their practice at time of interview, constitutes the absolute archetype of glocalisation.

Reflections

An indication of various occupational artefacts that might be valued by chefs and souvenired, was provided earlier and the data revealed that many of these carried organisational, systemic, individual (Wood, 1966) and occupational meaning. A range of discourses emerged. As expected, much was revealed about the uniform and its various components, and hence this becomes the focus of discussion. The uniform was an attraction to the occupation itself. Brooke recalled, 'I supposed you first want to be a chef, it is the uniform. There is a little bit of something to it … authority'. The uniform was absolutely integral to occupational membership and inculcated during training. Dale related that she first learnt the trade 'in a beautiful school [from] people in uniforms so clean and "cool"' and Suzie 'missed the traditionalism of dressing in "whites" … that has been a pleasure'. The communal value of the uniform was apparent in the distain chefs had for those that did not respect it, like young apprentices who 'you see with their chef pants and their shirts and they are filthy and they are walking around like... they come into "Maccas" [McDonalds restaurants]. Just not very good for the image', bemoaned Ebby.

To have the right to wear a chefs' jacket conveyed a rite of passage discourse. Thomas explained, 'they slap a jacket on them and there you go – you are a chef! You are not qualified but because there is such a lack in the industry for qualified chefs, you get a monkey and put a jacket on them … a lot of "never was's" around too who put jackets on and claim to be chefs.'

Thomas wore his jacket with pride and presumably Charles still keeps his signature-embroided jacket hanging on the back of his office door, even though he has long since been 'off the tools', to authenticate his occupational rite of passage but also his individual status (Wood, 1966), as typically only executive chefs had their jackets embroidered (see Figure 4.2). Cath, long since retired from cookery, still keeps her 'special selection of lucky aprons'. When Stan left his last ever chef job he recalled receiving, as a farewell gift, an old tattered jacket, signed in permanent marker pen by the kitchen brigade, picture framed with a note 'break in case of emergency'. Unfortunately, the memento 'broke' when his family moved home some years later! Perhaps Ebby too, talking about life after cooking, has spoken about a business partnership with her father, 'a warehouse sort of thing, like tools and uniforms' so she still has the occupational artefacts in her life. As with previous studies (Palmer *et al.*, 2010) the chefs clothing conveyed strong community membership, so it is unsurprising chefs valued their occupational apparel, but also retained them beyond retirement.

On the other hand those in positions of authority in the kitchen demonstrated their power by 'mix and matching'. A Michelin star chef headed

Figure 4.2 Chefs pose for marketing collateral

up a kitchen were Matt worked. 'I have total respect for the guy. We used to bag [tease] him because he used to work in a pair of jeans sometimes and he goes "the jeans do not make me cook any better"'. Rowdy, who fancied himself as a ladies' man, routinely wore black trousers instead of the regular checks to assert his status – 'the girls go wild on that authority, don't they?' he added. Enough of the chefs' uniform was kept in these two cases to assert 'occupational jurisdiction' (Bechky, 2003), an interesting observation given the various colourful and stylistic variations now worn by chefs (Sockrider, 2005) that vary from the traditional uniform (shown in Figures 4.1 and 4.2). There was less discussion about the *toque blanche*. There were suggestions that it was an artefact symbolic of status especially in more upmarket establishments where 'they will require you [to] have to wear a certain hat or whatever' related Seb, as he motioned with his hands above his head. Retired Stan still keeps his cloth *toque blanche*, neatly wedged between a couple aprons of sentimental value with a hardly worn 'special occasion jacket' in a cupboard.

While the scope of this chapter has focused mainly on the occupational experiences of chefs in an Australian context, several factors suggest a broader relevance of occupational artefacts to their occupational culture, especially as a shared global phenomenon. Two broad themes, in particular, were emergent from this study. First, there is evidence to suggest that the chefs uniform, and its various components, are highly symbolic as occupational artefacts, while also naturally of professional utility. Second, collecting behaviours suggest that these occupational artefacts assume the status of souvenirs even, should they become redundant through career change or retirement.

As occupational artefacts the uniform conveys membership rhetoric. The uniform, and its adjuncts, like the *toque* and aprons, communicated legitimacy – Thomas's indignity at 'monkeys' masquerading in jackets and Charles' signature-embroidered jacket hanging in his office being cases in point. Taking licence with the uniform, for example wearing black trousers instead of 'checks', was a marker of authority and identified status within a kitchen hierarchy. Collectively, these occupational artefacts made the chef 'king of the domain', the domain being the kitchen and king being an occupational jurisdiction (Bechky, 2003), so souveniring these artefacts provided the chefs with tangible evidence of their inclusion to a traditional occupation. More than just of nostalgic value, occupational souvenirs are reminders of social and economic 'usefulness', which in both a shared occupational sense and a tourism context transcends the local, Cath's 'lucky aprons' being a case in point. On the other hand Stan's souveniring of his toque is a reminder of occupational uniqueness, as distinct from a range of other occupations, a tangible artefact demonstrative of Goffman's (1959) back-stage-front stage performativity.

Conclusions

This chapter aimed to focus on, first suggesting the salience of a range of occupational artefacts to the community and culture of chefs, and second, to find evidence of the import of the uniform to identity and souveniring behaviours. In particular this study set out to explore how chefs' occupational souvenirs interconnect with their rich occupational culture and sense of community (e.g. Lee-Ross, 2005) and interpret souvenirs and the process of souveniring itself as symbolic of various dimensions of the occupational culture of chefs. These include the very nature of community, identity and authenticating membership (in time and place), marginality and deviance, mobility and the human resource management challenges of chefs as manifest through the symbolism of occupational souvenirs and finally souvenirs as expressive of the intrinsic rewards sought by chefs in their occupational experience.

Unlike the gift acquiring and giving motivations of tourists (Wilkins, 2011) chefs appear to cherish and hoard their uniforms and its parts such that they become symbols of membership and objects of reminiscence, which is in stark contrast to the uniform's mundane occupational function. Hence, these occupational artefacts and souvenirs communicate in a language that the 'Iron Chefs' of the Orient and their counterparts in Australia both understand – truly glocal occupational artefacts. But more than this they are understood not just by chefs themselves, but also tourists, be they food lovers or not, and via marketing and the media – glocal audiences too.

References

Ashforth, B.E. and Greiner, G.E. (1999) 'How can you do it?': Dirty work and the challenge of constructing a positive identity. *The Academy of Management Review* 24 (3), 413–434.

Bechky, B.A. (2003) Object lessons: Workplace artifacts as representations of occupational jurisdiction. *American Journal of Sociology* 109 (3), 720–752.

Carlile, P.R. (2002) A pragmatic view of knowledge and boundaries: Boundary objects in new product development. *Organization Science* 13, 442–55.

Elkjaer, B. and Huysman, M. (2008) Social worlds theory and the power of tension. In D. Barry and H. Hansen (eds) *The SAGE Handbook of New Approaches in Management and Organization* (pp. 170–177). London: Sage.

Fine, G.A. (1996) Justifying work: Occupational rhetorics as resources in restaurant kitchens. *Administrative Science Quarterly* 41 (1), 90–116.

Gallagher, M. (2004) What's so funny about iron chef? *Journal of Popular Film and Television* 31 (4), 176–184.

Goffman, E. (1959) *The Presentation of Self in Everyday Life*. New York: Anchor Books.

Gunders, J. (2008) Professionalism, place, and authenticity in 'The Cook and the Chef'. *Emotion, Space and Society* 1 (2), 119–126.

Harris, L.C. and Ogbonna, E. (2002) Exploring service sabotage: The antecedents, types and consequences of frontline, deviant, antiservice behaviors. *Journal of Service Research* 4 (3), 163–183.

Lashley, C., Morrison, A. and Randall, S. (2003) My most memorable meal ever: Some observations on the emotions of hospitality. In D. Sloan (ed.) *Culinary Taste*. Butterworth-Heinemann: Oxford.

Lupton, D. (1994) Food, memory and meaning: The symbolic and social nature of food events. *The Sociological Review* 42 (4), 664–685.

Lynch, O.H. (2009) Kitchen antics: The importance of humor and maintaining professionalism at work. *Journal of Applied Communication Research* 37 (4), 444–464.

Middleton, G. (2000) A preliminary study of chefs' attitudes and knowledge of healthy eating in Edinburgh's restaurants. *International Journal of Hospitality Management* 19 (4), 399–412.

Mullen, J. (1997) Homeless recipe for skill deficit. *People Management* 2 (23), 3–7.

Murphy, J. and Smith, S. (2009) Chefs and suppliers: An exploratory look at supply chain issues in an upscale restaurant alliance. *International Journal of Hospitality Management* 28 (2), 212–220.

Murray-Gibbons, R. and Gibbons, C. (2007) Occupational stress in the chef profession. *International Journal of Contemporary Hospitality Management* 19 (1), 32–42.

Okumus, B., Okumus, F. and McKercher, B. (2007) Incorporating local and international cuisines in the marketing of tourism destinations. *Tourism Management* 28 (1), 253–261.

Ottenbacher, M. and Harrington, R. (2013) A case study of a culinary tourism campaign in Germany: Implications for strategy making and successful implementation. *Journal of Hospitality and Tourism Research* 37 (1), 3–28.

Palmer, C., Cooper, J. and Burns, P. (2010) Culture, identity, and belonging in the 'culinary underbelly'. *International Journal of Culture, Tourism and Hospitality Research* 4 (4), 311–326.

Peterson, Y. and Birg, L. (1988) Top hat: The chef as creative occupation. *Free Inquiry in Creative Sociology* 16 (1), 67–72.

Robinson, R.N.S. (2008) Revisiting hospitality's marginal worker thesis: A mono-occupational perspective. *International Journal of Hospitality Management* 27 (3), 403–413.

Robinson, R.N.S. and Beesley, L.G. (2010) Linkages between creativity and intention to quit: An occupational study of chefs. *Tourism Management* 31 (6), 765–776.

Rowley, G. and Purcell, K. (2001) 'As cooks go, she went': Is labour churn inevitable? *International Journal of Hospitality Management* 20 (2), 163–185.

Sockrider, G.D. (2005) History of the chefs uniform, *Chefolio*. Pflugerville, Texas: Escoffier Media. Available from http://www.chefolio.com/Articles/HistoryoftheChefsuniform.html (accesed 29 June 2009).

Sparkes, B.A., Bowen, J. and Klag, S. (2003) Restaurants and the tourism market. *International Journal of Contemporary Hospitality Management* 15 (1), 6–14.

Tajfel, H. (1974) Social identity and intergroup behavior. *Social Science Information* 13 (1), 65–93.

The Culinary Institute of America (2001) The chef's uniform, *Gastronomica. The Journal of Food and Culture* 1 (1), 88–91.

Turner, J.C. (1975) Social comparisons and social identity: Some prospects for inter-group behavior. *European Journal of Social Psychology* 5, 5–34.

White, A., Jones, E. and James, D. (2005) There's a nasty smell in the kitchen! Gender and power in the tourism workplace in Wales. *Tourism, Culture and Communication* 6, 37–49.

Wilkins, H. (2011) Souvenirs: What and why we buy. *Journal of Travel Research* 50 (3), 239–247.

Wood, R.C. (1992) Deviants and misfits: Hotel and catering labour and the marginal worker thesis. *International Journal of Hospitality Management* 11 (3), 179–182.

Wood, R.C. (2000) Why are there so many celebrity chefs and cooks (and do we need them)? Culinary cultism and crassness on television and beyond. In R.C. Wood (ed.) *Strategic Questions in Food and Beverage Management* (pp. 129–152). Oxford: Butterworth Heinemann.

Wood, S.M. (1966) Uniform: Its significance as a factor in role-relationships. *The Sociological Review* 14 (2), 139–151.

WTTC (2011) Economic impact data and forecasts: World key facts at a glance. World Tourism and Travel Council (WTTC).

Part 2

Theorising Place and Identity

5 Souvenirs of the American Southwest: Objective or Constructive Authenticity?

Kristen K. Swanson

Apache, Hopi, Navajo, Pueblo and Zuni are among the 35 Native American tribes who live in the American Southwest (Bahti, 1992). Each population group is autonomous, with independent traditions, values, customs, artistic expression, religious practices, food, dress, tribal government structure and sense of self-concept constantly working to blend both traditional and modern ways of life as indigenous populations.

As with many indigenous cultures, these population groups have experienced marginalisation as a result of colonisation. The native peoples of the Southwest whose lands fall within the national boundaries of the United States are members of the fourth world (Graburn, 1976). They are often treated as a minority without complete power to direct the course of their lives. Each group has continually fought to exercise their cultural rights as independent sovereign nations. Cultural rights emphasise people's awareness of their cultural identity, their right to be different and of the mutual respect of one culture for another (Symonides, 1998). However, as a collective group of Native American tribes in the Southwest, they share many of the same challenges as marginalised populations within first-world societies. Conditions of educational disadvantage, political disenfranchisement, exploitative relationships and unemployment threaten many communities. Geographic remoteness, cultural values, and political and religious factors restrict opportunities for economic development. Economic exclusion is at the heart of the problem of marginalisation (Bhalla & Lapeyre, 1997).

Throughout history the cultures of the Southwest have retained centuries-old traditions that form their sense of place, while at the same time accepting change when consistent with their values and world views

(Whiteford *et al.*, 1989). The sense of place, way of life and worldview dominated in these cultures provides localisation (Salazar, 2005), while at the same time, the necessity for these cultures to participate within the realms of the first-world affords, if not requires, globalisation. In the Southwest, souvenirs function as representations of place and people that have occupied the region since ancient times and simultaneously function as tradable commodities used by Native American populations for centuries as a main source of income, for survival and livelihood. Using souvenirs as the subject of study, the aim of this chapter is to explore the marginalisation of the objectively authentic American Southwest as a consequence of occupation by non-indigenous populations. Interestingly, the souvenirs that most represent the objectively authentic American Southwest in modern times are really constructed authentic mainstream representations of the ancient indigenous cultures of the area – further marginalising the original, authentic and real.

For over two centuries the arts and crafts of the American Southwest, produced by indigenous peoples of the area, have been collected, first for curiosity value and later for aesthetic value (Graburn, 1976). This has allowed the production and distribution of local, indigenous art forms as tourist souvenirs to be seen as a potential avenue for enhancing economic development. Capitalising on local cultural resources allows rural producers, such as those in the American Southwest, to survive and retain control locally, in a growing international society (Cawley *et al.*, 2002). However, it has also had the effect of turning the cultures into commodities to be sold to tourists, resulting in the loss of authenticity (Cole, 2007). The basis for this work is formed through recent case study research (Swanson & DeVereaux, 2012), anecdotal evidence and prior studies focused on souvenir phenomena in the American Southwest (Swanson, 1994, 2004; Swanson & Horridge, 2002, 2004, 2006).

Place and People

Souvenirs representing the American Southwest function as messengers that give meaning to a place – roughly the 'four corners' area of the United States – and as messengers giving meaning to the Native American populations of the region. The Four Corners is a region of the United States consisting of the southwestern corner of Colorado, northwestern corner of New Mexico, northeastern corner of Arizona and southeastern corner of Utah, part of a larger region known as the Colorado Plateau that is mostly rural, rugged and arid. The literal meaning indicates the only place in the United States where four states intersect at one point. However, a more constructed interpretation of the 'four corners' region indicates a geographic region that is part of the semi-autonomous indigenous American Indian nations.

The American Southwest is an example of a natural and cultural destination that has been brought to the global stage as a result of increased population mobility and technical advances (Urry, 2000), and the growing international travel and tourism industry (Jamal & Hill, 2004). Visitors travel to the area to experience the living and past culture (Budruk et al., 2008). Tourism in the Southwest offers the 'Paradoxical promise: a self-fulfilling, one-of-a-kind experience, as a mass produced phenomenon' (Shaffer, 2003: 73). It is the sunniest and driest region in the US. It includes high mountain ranges as well as arid deserts, and broad flatlands and plateaus. The Rio Grande and Colorado Rivers are part of the Southwest as well as the Grand Canyon and Canyon de Chelly. The Southwest is often visualised through dramatic pictures of mesas and buttes with a clear, blue sky in the background. Both clear and muted colours of brown, tan, blue, green, yellow, orange, red, purple and pink symbolise the Southwest landscape. The colours of the Southwest were originally created by Native Americans who mixed paints and dyes from local plants. The colours are inspired by landscape, and all forms of nature – desert, sky, water, plants and sunshine.

The American Southwest has always carried with it a mystic quality inspiring a sense of mystery and wonder. The message on a billboard along Interstate 40 between Arizona and New Mexico invites drivers to visit 'mystic Gallop'. The sign could as easily read, 'visit "mystic Kayenta", "mystic Belen", or "mystic Española"'. The Southwest has long held a degree of fascination that appealed to the 'frontier spirit of the American personality' (Deitch, 1989: 226). Southwest tourist souvenirs are uniquely expressive and distinguishable to this region of the world. In research on the impact of tourism on the arts and crafts of the Southwest Indians, Deitch (1989) said, 'throughout the nation, people are aware of the native arts of the Southwest, and many are eager to possess something Indian' (p. 224). He continued that, 'for the first time since white contact, the Indian has a commodity that is sought by members of Western society'. In a study of Canyon de Chelly, located on the Navajo reservation, a strong sense of place was evident, and having a strong sense of place contributed to a genuine or authentic experience at that place (Budruk et al., 2008). In discussing Southwest souvenirs, Dilworth (2003) considered the exchange between Indian artisans and tourists not to be a 'real market', but rather a 'magical exchange, in which a mystical essence that has rubbed off on the object from the caress of an Indian hand can be accumulated by the buyer' (p. 112).

The American Southwest is a complex blend of Pueblo, Athabascan, Spanish and Anglo cultures. The current populations are descendants of cultures who inhibited the region in ancient times (i.e. Pueblo, Yavapai, Havasupai, Pima, Tohono O'odham) and who arrived in the late 15th and early 16th centuries (Navajo, Apache) (Plog, 2008). The first inhabitants, 10,000 BC, were descendants of immigrants from Asia who crossed the Bering Straits land bridge (Whiteford *et al.*, 1989). By 100 BC, these groups had developed farming villages and divided into four major cultural groups: Hakataya, Hohokam, Anasazi and Mogollon. The dissemination of culture through material objects was evident as early as 500 AD, when Mogollon and Anasazi cultures began manufacturing and trading pottery, which replaced baskets for storing food. Coiled basket weaving, represented in present-day souvenirs, is an example of a technique used by the earliest dwellers.

By 900 AD, smaller Anasazi communities had dispersed over most of the Colorado Plateau while larger communities appeared in the San Juan region of New Mexico. These cultures collapsed during the late 13th century for reasons that are still not fully understood. Following the collapse of the larger communities, large pueblo populations developed along the Río Grande River, the Little Colorado River, and on the three Hopi mesas. By this time the Pueblo people were proficient in the creation of pottery, baskets, cotton cloth, shell jewellery and kachina figures (Deitch, 1989).

Athabasca people, ancestors of present day Apache and Navajo cultures, migrated to the Southwest from Western Canada in the 15th century (Whiteford *et al.*, 1989). They did not have the artistic sophistication of the Pueblo dwellers, but were quick to learn weaving techniques from the Pueblo cultures (Deitch, 1989). The combination of weaving and wool (adopted when the Spanish brought sheep to the area in later times) became an integral part of Navajo artistic expression. Today, this adaption between cultures is symbolised in highly valued Navajo rugs and other weavings.

Spanish explorers discovered the Southwest in the 16th century. European influence was introduced by conquistadors – soldiers, explorers, and adventurers at the service of the Spanish or Portuguese Empire–who brought cattle, sheep, soldiers, Mexican Indians and priests. Extreme exploitative, marginalised relationships came about almost immediately when Juan de Oñate proclaimed the Pueblo people subjects of the Spanish Crown and the Catholic Church (Weinstein, 2002). In times of peace the Pueblo, Apache and Navajo, and Spaniards exchanged ideas, material goods and food (Whiteford *et al.*, 1989). Exchange led to new designs, techniques and materials in art forms; a process that continues in present time. Silverwork was diffused into the culture by Mexicans who came with the Spaniards. Silver jewellery was first obtained by Navajos in trade for horses, but soon they manufactured their own silver ornaments (Deitch, 1989). The squash-blossom

Figure 5.1 A traditional Navajo squash-blossom necklace

necklace, a 'typically Indian' design (p. 225) is a Navajo interpretation of a Spanish pomegranate design (see Figure 5.1). The Navajo diffused silverwork techniques to Zuni and Hopi. The turquoise stone, considered an iconic image of the American Southwest, was first mined in the area at the end of the 19th century but was quickly dispersed into the Southwest culture. In present times, high-end and curio, turquoise and silver jewellery from Navajo, Zuni and Hopi artisans are popular tourist souvenirs.

Many Native American populations continue to find beauty in the place of the American Southwest, using this beauty as aesthetic inspiration for their art. Their lives are centred on living life in concert with the values of humility, cooperation, respect and universal earth stewardship and these values are reflected in weavings, paintings, sculpture and other artistic forms.

Souvenirs and Authenticity

Souvenirs are objects that evoke memories of an experience. Individuals often consider souvenirs to be among their most prised possessions (Littrell, 1990). A number of tourism studies have expressed the relationships between the souvenir object, place visited and person as visitor (Swanson & Timothy,

2012). The object–place relationship recognises that the souvenir allows the visitor to make sense of the visit during and after the experience. During travel the souvenir object serves to heighten the experience of travel; after returning home the souvenir serves as a holder in time of that past experience (Gordon, 1986). The souvenir functions as proof of travel and allows reflection of the sacred journey of travel (Graburn, 1989). The souvenir is symbolic of some impression of the Southwest (i.e. geography, materials, artistic representation, historical reference, elements of place) while the visitor is in the region as well as after they return home.

In the place–person relationship, the souvenir is representative of a specific place (i.e. Hopi Mesa, Santo Domingo Pueblo, Ganado Trading Post). The souvenir captures the unique qualities of the destination and transports those qualities home as reminders of what made the place special (Swanson & Timothy, 2012). Years after the trip, visitors are still immediately transported back to the place, at mention of a particular souvenir. The geographic scale of the souvenir may represent a country, region, city, specific attraction or a combination of several geographical scales (Hashimoto & Telfer, 2007). In the Southwest, tourist souvenirs may represent America (Pendleton blanket); the Southwest (moccasins); a community (Jemez storyteller); or specific attraction (replica dwelling from Mesa Verde). Each souvenir holds unique meaning for the possessor. A variety of souvenirs representing all geographical scales is evident throughout the Southwest, potentially capturing the unique qualities of both the destination and the people who live there.

A myriad of objects comprise Southwestern tourism souvenirs. The artistic expression of its various Native American, Hispanic and Anglo cultures are enchanting, reflected in pottery, weavings, jewellery, *kachina* carvings and other forms that have evolved as 'authentic' symbols of the area. A sampling of the major art forms is briefly described here. Each art piece is created to express the individuality of the artist and perhaps the possessor of the tourist souvenir (Simpson, 1999).

Pottery is one of the oldest art forms dating back two thousand years (Lamb, 1996). Making pottery is an honoured tradition and still plays a role in ancient ceremonies. The pottery from each pueblo has a distinct appearance owing to features of the clay (i.e. Taos, Nambé); colour (i.e. Tesuque) and design aesthetics (i.e. Ácoma, Santa Clara) of the specific community (see Figure 5.2). Most potters are women and some families have made fine pottery for generations.

The Navajo people adopted *weaving* from Pueblo communities in the 1600s. However, commercial yarns, dyes and colours, together with railroad clientele, significantly changed the art form (Whiteford *et al.*, 1989). By the late 1880s, hundreds of blankets were woven to meet demand, and

Figure 5.2 Southwestern pots representing Santa Clara pottery (black) and Hopi pottery (brown)

interestingly, the Navajo themselves stopped weaving for home use and instead started purchasing Pendleton blankets (which too have become a subjectively authentic representation of the American Southwest.) Navajo rug styles are distinguished by colour alone (i.e. Ganado, Two Grey Hills); bands of colour and band designs (i.e. Chinle, Wide Ruins); or by design alone (i.e. Teec Nos Pos, Storm, shown in Figure 5.3) (Lamb, 1992).

Jewellery has been a part of the dress of nearly every Native American culture in the US (Baxter & Bird-Romero, 2000). American Indian jewellery symbolises the special relationship between humans and the natural and spiritual worlds, and is a type of spiritual expression and means of cultural preservation (Schaaf, 2003). Jewellery is distinguished by the people crafting the objects, and materials and forms (i.e. Pueblo fetish jewellery); techniques (i.e. Navajo castwork, Zuni mosaic inlay); and design motifs (i.e. Hopi overlay, shown in Figure 5.4, O'otam maze design).

Kachina is specific to Pueblo Indian religions (most prevalent among Hopi and Zuni) and refers to spiritual beings and dolls, more accurately used in the Hopi language as *katsina* (Secakuku, 1995), and *koko* in the Zuni language (Bahti, 1992). The word refers to the spirit being, applying to the male dancers who impersonate the spirit beings, and painted wooden figurines that symbolise the masked dancers (see Figure 5.5). Kachinas are very important in the religious life of Pueblo people and are depicted in nearly 200 variations representing the spirits of birds, animals (i.e. Bear kachina), places, objects, forces of nature (i.e. Snow maiden), insects, plants and other tribes. Pueblo men carve kachinas from the cottonwood tree root and

Figure 5.3 A storm pattern Navajo rug

Figure 5.4 Hopi inlay design on a bracelet

present the carvings as blessings to children, young girls and women during a kachina dance.

As in the case with many indigenous cultures, the artisan objects considered as tourist souvenirs in modern society were originally produced to fulfil utilitarian needs. Baskets and pots were made for cooking and storage; textiles were used for clothing and shelter. Other items were made as religious

Figure 5.5 A Hopi kachina carving

symbols to honour or serve as offerings to gods (Swanson & Timothy, 2012). In more recent times, these items have been transformed into souvenir commodities for the tourism market. For example, Pueblo ceramic pots, first used to make food storage, cooking and serving containers, were transformed into handmade marketable commodities to sell to white tourists for cash (Brody, 1979). The unsophisticated customers wanted small exotic pots and were not concerned with the craft qualities, giving way to 'badly made, badly fired, cheap, exotic articles' (p. 74).

Since MacCannell (1976) introduced the concept of authenticity in tourism, much has been written about the subject. Authenticity has elements of both object–person and object–place relationships. Some scholars contend authenticity is real and can be measured; objects and places are inherently authentic, and genuineness of the object may be quantified (Reisinger & Steiner, 2006; Wang, 1999). Authentic crafts as handmade items made with attention to materials, design and workmanship (Littrell *et al.*, 1993). The knowledge that a cultural souvenir is locally handmade is important in establishing authenticity (Healy, 1994). Objective authenticity, supports native producers rights' that items sold can be handcrafted art forms, individually made by native artists of the culture, genuine because of the legitimacy of the person crafting the item to represent his or her culture. A hallmark, a silversmith's personal symbol stamped on the back of Native American jewellery (Wright, 1989) is meant to represent quantifiable objective authenticity.

However, other scholars content that authenticity is subjective, based on the tourist's connection with the object constructed from preconceived notions or cultural biases (Budruk *et al.*, 2008). Subjective authenticity does not have to be a concrete characteristic of the object; rather, it can be based on the individual's connection with the object, the purchase experience, place or other idiosyncratic characteristics (Budruk *et al.*, 2008). For example, this author owns an objectively authentic squash-blossom necklace, handmade by a native producer and purchased directly from the artisan (Figure 5.1). Additionally she owns a T-shirt, embossed to imitate a squash-blossom necklace (see Figure 5.6), and produced by a non-native manufacturer, representing subjective authenticity. For this researcher/tourist, both objective and subjective authentic items provide connectedness to the place and people of the American Southwest. Using the hallmark example, non-native producers, realising that tourists may be uneducated or unconcerned

Figure 5.6 Subjective authenticity: Embossed representation of a squash-blossom necklace

about the genuineness of an object, have created fake hallmarks underscoring the marginalisation of the Indian artisan, implying the craft is 'Indian made' when it is really produced in Asia or Mexico (Brooke, 1997; Parsley, 1993).

Another perspective of authenticity relates to constructive authenticity in which things appear authentic, not because they are inherently authentic, but because they have been constructed as such in terms of point of view, beliefs, perspectives and power (Wang, 1999). In the Southwest, the constructed point of view has often come at the hands of non-native producers. For example, Taos painters of the 1950s depicted 'archetypical and idealized Indians' (Swentzell, 2003: 66). However, Native Americans who lived in Taos did not recognise the 'Indians' of the painting; the paintings caused the Pueblo people to feel unworthy, not measuring up to how they were represented (Swentzell, 2003).

Another constructed view of the American Southwest was exploited by the Fred Harvey Company and the Santa Fe Railroad. The Santa Fe/Harvey corporations appropriated, displayed and marketed the cultures of Native American, Spanish colonial and Anglo hunter/trapper/prospectors to create a compelling regional identify for 'The Great Southwest' (Weigle, 1989: 115). For advertising purposes, the railway purchased paintings of Southwest images including *Grand Canyon* by Thomas Moran and *El Tovar* by Louis Akin. Thousands of duplicate prints were made of these art works and displayed across the United States, placed in railroad stations to lure tourists to visit the American West to see unparalleled scenic views, participate in high adventures, experience romantic journeys, meet Native American people and observe their culture (Howard & Pardue, 1996). At train stops Native Americans would gather to sell hand-woven blankets, baskets, silver jewellery and pottery, which tourists quickly bought up. Harvey recognised the marketability of Navajo rugs, jewellery, pottery, basketry, beads and kachinas as tourist souvenirs, 'To the visitor, this was the epitome of the Southwest, and nearly everyone bought something to take home' (Deitch, 1989: 226). He opened the Fred Harvey Company curio business in 1899, as a non-native producer, and masterfully constructed 'authentic' Indian art forms, into a significant commercial souvenir market.

Native Americans have continually remade their arts and crafts to meet the needs and tastes of non-native tourists, compromising objective authenticity for constructive authenticity. Phillips (1998) described this as intercultural negotiation. The artisan has to continually translate their culture through small adjustments to produce souvenir arts for the non-native market that balance authenticity with modernity concerns of both tourist and producer. The artisan continually tailors local products that represent his or her heritage and culture to changing global audiences (Salazar, 2005).

As defined above, objectively authentic and constructively authentic souvenirs co-exist in the marketplace. For example, at one well-known historic trading post, objectively authentic silver and turquoise jewellery is shown in enclosed glass display cases, in some instances costing hundreds of dollars, adjacent to open fixture display racks with mass-produced, constructively authentic imported jewellery of similar style and design selling for tens of dollars. In another example, an art gallery sells objectively authentic, hand sewn accessories made from Pendleton blankets by native producers located adjacent to subjectively authentic Pendleton imprinted, mass-produced coffee mugs, from non-native vendors.

Historically, the tourist influence on native-made objects was both positive and negative (Baxter & Bird-Romero, 2000). From a positive perspective, tourist souvenirs provided an essential income for Native Americans. In addition, the demand for tourist souvenirs allowed designs, that would have otherwise died out, to be modified and revived (Brody, 1979). From a negative perspective, the lower quality Indian items, particularly jewellery, made for the non-native tourist market representing constructive authenticity harmed the objectively authentic market, a problem that still persists in present day.

Issues

Native American cultures within the American Southwest are at the centre of cultural tourism in the region. These population groups continue to be challenged by several issues. First, tourism is invasive for indigenous cultures since many are living communities in which the day-to-day native setting of the host is also the tourist attraction for the guest. 'The local people themselves and their way of life become the focus of the tourist "gaze" (Timothy, 2011: 425). It is not uncommon for Pueblo artisans to sell pottery directly from the homes. Indigenous residents recognise the potential benefits of employment and economic gain through tourism, but do not want it to disrupt the safety and well-being of tribal members (physical as well as cultural) or sacrifice their traditional ways of life (Swanson & DeVereaux, 2012). At one extreme, some Native American groups believe that for the cultural well-being of the community, only objectively authentic products made by Native Americans should be sold. Other groups consider the economic well-being of the community to be of greater importance and are willing to sell, constructively authentic (imported) souvenirs for economic gain as long as all income earned remains with the Native American community. One Native American informant stated that her tribe would rather sell fewer locally made, authentic carvings, at a price point well above the typical tourist's ability to pay, rather than countless, more affordable but mass-produced imports, even forgoing economic gain.

The second issue facing Native American groups is ownership and identification of what is inherently their own. Non-native producers are manufacturing and selling tourist souvenirs with no input from the originating Native American culture, made by people who have little understanding or training in the traditional artisan skills. Objective authenticity is dubious as non-local materials are used, non-original techniques are adopted for ease of production and design details are ignored, creating inaccurate and stereotypical perceptions of the Southwest and the Native American population groups. For example, in one community all vendors are supposed to be licensed through a permitting process as established by a tourism policy. However, unless the area is constantly patrolled, squatters pretending to be members of the community sell non-native products to uninformed tourists who believe they are making legitimate contributions to the host community. Issues arise concerning claims to origin, originality, materials and manufacturing processes when local indigenous groups are not involved (Asplet & Cooper, 2000). Academics and natives rights groups have declared these exploitative relationships unfair and illegal (Blundell, 1993; Brooke, 1997), but have not been able to legislate cultural property rights in a meaningful way regarding the reproduction and selling of indigenous arts, crafts and symbols by non-native producers, even with the Indian Arts and Crafts Act of 1990 (Guest, 1995/1996; Parsley, 1993).

Continual modification of ancestral forms to meet the changing demands of non-native tourists who are continually negotiating their subjective notion of authenticity is a concern to for Native Americans. Because the Native American cultures of the Southwest are autonomous, some groups have willingly however changed their culturally unique artisan expressions to meet the demand of non-native tourists who are willing to pay for representations of ancestral forms. One common example is Christmas ornaments. The Christian festival marking the birth of Jesus Christ is not an ancestral religious practice of many indigenous cultures, yet many groups produce ornaments with native designs to sell to non-native tourists (see Figure 5.7). Through the commodification process, the souvenir loses its cultural meaning and further marginalises the objective authentic. Modification of culturally unique, ancestral art forms to satisfy the non-native tourist market, have resulted in several negative effects (Turco, 1999: 57): such as: (1) the loss of meaning of cultural traditions and arts as they become commoditised; (2) the change in size, structure and colour of artefacts to attract a buying public; (3) a change from artisan intimacy with products to artisan disinterest

Figure 5.7 Constructed authenticity: Christmas ornaments including a representation of Zuni design, Pueblo pot and Snowmaiden kachina

because of mass production; (4) garish billboards and enlarged figurines to attract passers-by and (5) encroachment upon sacred and culturally meaningful ceremonies and lands by visitors.

Another marginalisation issue for Native Americans is the persistence of the Hollywood versions of the colonisation of the American Southwest as stereotyped cowboys and Indians (Plog, 2008). Native Americans must weigh the potential for increased economic prosperity against the cost to the cultural rights as a sovereign people. Often times, producing souvenirs means producing something constructively authentic and stereotyped as 'Indian'. While the Santa Fe/Harvey corporations marketed the Native American cultures, they also developed exploitative relationships with the region's native peoples, popularising the concept of 'Indianness' at the turn of the 20th century. Native Americans were depicted as primitive and put on display at the World's Fair, (Weigle, 1989), and as objects of staged authenticity (MacCannell, 1999) through Indian Detours, chauffeured, guided car excursions that enabled rail tourists to see points of historic and scenic interest, including visits to Pueblo Indian home life and crafts (Weigle, 1996). The Indianness concept is still evident in the Southwest in present day. Indian warrior dolls, complete with feathered headdress and tomahawk, can be found for sale throughout the Southwest. In present day, a billboard invites travellers to visit *Indian City, Navajo-owned* with the slogan, *'Smile and Say Chee's'*. As drivers pass *Chief Yellow Horse*, a roadside Indian-owned business, they are encouraged to turn around and see *'Friendly Indians behind you'*. Because cultural and economic sustainability for Native American populations of the Southwest has always been a concern (Cornell & Kalt, 1992; Rosser, 2005), Native Americans themselves are promoting the stereotypical primitive Indian of the Fred Harvey era, compromising their cultural identity to potentially lure a non-native customer. For many, economic stability including souvenir production and marketing has come at the price of compromise of cultural traditions and values.

A last obstacle facing Native American artisans is the disinterest of current generations in learning the crafts and preserving the cultural art forms, leading to further marginalisation (Parsley, 1993). For example, at one historic trading post a Navajo weaver produces beautiful rugs that sell for thousands of dollars. However, none of her children are interested in learning weaving and her craft will not be passed on.

Final Thoughts

Souvenirs of the American Southwest represent a diffusion of interwoven cultures based on centuries of acculturation. With the introduction of each new culture, the art forms have changed, producing new images displayed in material culture. Additionally, new cultures have created exploitative relationships that have marginalised Native Americans and threatened traditional ways of life. The past cannot be changed. However, going forward, echoing Cornell and Kalt (1992), indigenous cultures including Native Americans need to exercise their sovereign right as the intellectual owners of their material cultures and decide what to sell to tourists. Indigenous populations should continue to promote and exhibit cultural rights in the representations of souvenirs of the culture. Additional research and application of global legislative tools, such as the Indian Arts and Crafts Act of 1990 in the US, may be a starting point.

Marginalisation is not easily reversed. Future research should focus on developing culturally sustainable entrepreneurial opportunities that can lessen the economic exclusion of indigenous populations including Native Americans without compromise to cultural values (Swanson & DeVereaux, 2012). Some cultures are not comfortable with the concept of a market economy, and efforts should be made to increase this confidence (Cave, 2009).

The points of view presented in this chapter are not conclusive but rather suggestive and many opportunities exist for further research. However, it is imperative that participation and commitment to achieving a shared goal from the indigenous communities be respected, expected and accepted during the research process. With respect to future research, it is important to develop relationships that allow cultural rights and representations of culture to be explored without compromising the value systems or cultural well-being of the studied populations. For example, future research should focus on ways to lessen stereotyped images of indigenous cultures in souvenirs, such as the Hollywood image of the 'Indian', and develop innovative, original ways in which to represent contemporary indigenous people and places.

Opportunities are ripe for future research projects focused on the local. Research studies have examined Native American issues from the non-native perspective, as outsiders looking in; future research should focus on gleaning information from the native perspective within the indigenous community. Additionally, although this chapter has looked at Native American souvenirs of the American Southwest with a broad stroke, future researchers should be cognizant of the autonomous nature of member groups within indigenous cultures and respect and treat each as independent cultures with their own set of values, customs, ways of artistic expressions, approach to objective and constructive authenticity, and issues related to marginalisation. It is unreasonable to believe that all souvenirs available in the American Southwest will be inherently objectively authentic. Therefore, too, it is in the best interest of indigenous cultures, including Native American cultures, to address subjective authenticity and determine what connections tourists make with objects, the purchase experience and the place. More research is necessary to understand objective, subjective and constructive authenticity with relation to souvenirs of the American Southwest.

References

Asplet, M. and Cooper, M. (2000) Cultural designs on New Zealand souvenir clothing: The question of authenticity. *Tourism Management* 21, 307–312.

Bahti, M. (1992) *Southwestern Indian Arts and Crafts*. Wickenburg, AZ: KC publications.

Baxter, P.A. and Bird-Romero, A. (2000) *Encyclopaedia of Native American Jewellery*. Phoenix, AZ: Oryx Press.

Bhalla, A. and Lapeyre, F. (1997) Social exclusion: Towards an analytical and operational framework. *Development and Change* 28 (3), 413–433.

Blundell, V. (1993) Aboriginal empowerment and souvenir trade in Canada. *Annals of Tourism Research* 20, 64–87.

Brody, J.J. (1979) The creative consumer: Survival, revival, and invention in the Southwest Indian Arts. In N.H.H. Graburn (ed.) *Ethnic and Tourist Arts* (pp. 70–84). Berkley: University of California Press.

Brooke, J. (1997, August 2) Sales of Indian crafts rise and so do fakes. *New York Times*. Retrieved http://nytimes.com.

Budruk, M., White, D.D., Wodrich, J.A. and van Riper, C.J. (2008) Connecting visitors to people and place: Visitors' perceptions of authenticity at Canyon de Chelly National Monument, Arizona. *Journal of Heritage Tourism* 3, 185–202.

Cave, J. (2009) Embedded identity: Pacific Islanders, cultural economies, and migrant tourism product. *Tourism, Culture & Communication* 9, 65–77.

Cawley, M., Gaffey, S. and Gillmor, D.A. (2002) Localization and global reach in rural tourism: Irish evidence. *Tourism Studies* 2 (1), 63–86.

Cole, S. (2007) Beyond authenticity and commodification. *Annals of Tourism Research* 34 (4), 943–960.

Cornell, S. and Kalt, J.P. (1992) *What Can Tribes Do? Strategies and Institutions in American Indian Economic Development*. Los Angeles: University of California, American Indian Studies Center.

Deitch, L.I. (1989) The impact of tourism on the arts and crafts of the Indians of the Southwestern United States. In V.L. Smith (ed.) *Hosts and Guests: The Anthropology of Tourism* (pp. 223–235). Philadelphia: University of Pennsylvania Press.

Dilworth, L. (2003) "Handmade by an American Indian" souvenirs and the cultural economy of Southwestern tourism. In H.K. Rothman (ed.) *The Culture of Tourism, the Tourism of Culture: Selling the Past to the Present in the American Southwest* (pp. 101–117). Albuquerque, NM: University of New Mexico Press.

Gordon, B. (1986) The souvenir: Messenger of the extraordinary. *Journal of Popular Culture* 20 (3), 135–146.

Graburn, N.H.H. (1976) Introduction: Arts of the Fourth World. In N.H.H. Graburn (ed.) *Ethnic and Tourist Arts* (pp. 1–32). Berkley: University of California Press.

Graburn, N.H.H. (1989) Tourism: The sacred journey. In V.L. Smith (ed.) *Hosts and Guests: The Anthropology of Tourism* (pp. 21–36). Philadelphia: University of Pennsylvania Press.

Guest, R.A. (1995/1996) Intellectual property rights and Native American Tribes. *American Indian Law Review* 20 (1), 111–139.

Hashimoto, A. and Telfer, D.J. (2007) Geographical representations embedded within souvenirs in Niagara: The case of geographically displaced authenticity. *Tourism Geographies* 9 (2), 191–217.

Healy, R.G. (1994) 'Tourist merchandise' as a means of generating local benefits for ecotourism. *Journal of Sustainable Tourism* 2 (3), 137–151.

Howard, K.J. and Pardue, D.F. (1996) *Inventing the Southwest: The Fred Harvey Company and Native American art*. Flagstaff, AZ: Northland Publishing.

Jamal, T. and Hill, S. (2004) Developing a framework for indicators of authenticity: The place and space of cultural and heritage tourism. *Asia Pacific Journal of Tourism Research* 9 (4), 333–371.

Lamb, S. (1992) *Guide to Navajo Rugs*. Tucson, AZ: Southwest Parks and Monuments Association.

Lamb, S. (1996) *Guide to Pueblo Pottery*. Tucson, AZ: Southwest Parks and Monuments Association.

Littrell, M.A. (1990) Symbolic significance of textile crafts for tourists. *Annals of Tourism Research* 17, 228–245.

Littrell, M.A., Anderson, L.F. and Brown, P.J. (1993) What makes a craft souvenir authentic? *Annals of Tourism Research* 20, 197–215.

MacCannell, D. (1976) *The Tourist: A New Theory of the Leisure Class*. Berkeley, CA: University of California Press.

Parsley, J.K. (1993) Regulation of counterfeit Indian arts and crafts: An analysis of the Indian Arts and Crafts Act of 1990. *American Indian Law Review* 18 (2), 487–514.

Phillips, R.B. (1998) *Trading Identities: The Souvenir in Native North American Art from the Northeast, 1700–1900*. Seattle: University of Washington Press.

Plog, S. (2008) *Ancient peoples of the American Southwest* (2nd edn). London: Thames and Hudson.

Reisinger, Y. and Steiner, C.J. (2006) Reconceptualizing object authenticity. *Annals of Tourism Research* 33, 65–86.

Rosser, E. (2005) This land is my land, this land is your land: Markets and institutions for economic development on Native American reservations. *Arizona Law Review* 47, 245–316. Retrieved from http://works.bepress.com/ezra_rosser/14

Salazar, N.B. (2005) Tourism and glocalization: 'Local' tour guiding. *Annals of Tourism Research* 32, 628–646.

Schaaf, G. (2003) *American Indian Jewelry I*. Santa Fe, NM: CIAC [Center for Indigenous Arts & Cultures Press].

Secakuku, A.H. (1995) *Following the Sun and Moon*. Phoenix, AZ: Heard Museum.

Shaffer, M.S. (2003) Playing American the Southwest scrapbooks of Mildred E. Baker. In H.K. Rothman (ed.) *The Culture of Tourism, the Tourism of Culture* (pp. 101–117). Albuquerque, NM: University of New Mexico Press.

Simpson, G.K. (1999) *Guide to Indian Jewelry of the Southwest*. Tucson, AZ: Southwest Parks and Monuments Association.

Swanson, K.K. (1994) Souvenir Marketing in Tourism Retailing: Shopper and Retailer Perceptions. Unpublished doctoral dissertation, Texas Tech University: Lubbock, TX.

Swanson, K.K. (2004) Tourists' and retailers' perceptions of souvenirs. *Journal of Vacation Marketing* 10 (4), 363–377.

Swanson, K.K. and DeVereaux, C. (2012) Culturally sustainable entrepreneurship: A Case study for Hopi tourism. In K. Hyde, C. Ryan and A.G. Woodside (eds) *Field Guide to Case Study Research in Tourism, Hospitality, and Leisure* (479–494). Bingley, UK: Emerald Group.

Swanson, K.K. and Horridge, P.E. (2002) Tourists' souvenir purchase behavior and retailers' awareness of tourists' purchase behavior in the Southwest. *Clothing and Textiles Research Journal* 20 (2), 62–76.

Swanson, K.K. and Horridge, P.E. (2004) A structural model for souvenir consumption, travel activities, and tourist demographics. *Journal of Travel Research* 42, 372–380.

Swanson, K.K. and Horridge, P.E. (2006) Travel motivations as souvenir purchase indicators. *Tourism Management* 27, 671–683.

Swanson, K.K. and Timothy, D.J. (2012) Souvenirs: Icons of meaning, commercialization and commoditization. *Tourism Management* 33, 489–499.

Swentzell, R. (2003) Anglo artists and the creation of Pueblo worlds. In H.K. Rothman (ed.) *The Culture of Tourism, the Tourism of Culture* (pp. 101–117). Albuquerque, NM: University of New Mexico Press.

Symonides, J. (1998) Cultural rights: A neglected category of human rights. *International Social Science Journal* 158, 559–573.

Timothy, D.J. (2011) *Cultural Heritage and Tourism: An Introduction*. Bristol: Channel View Publications.

Turco, D.M. (1999) Ya' 'at 'eeh: A profile of tourists to Navajo Nation. *Journal of Tourism Studies* 10, 57–61.

Urry, J. (2000) *Sociology Beyond Societies: Mobilities for the Twenty-First Century*. London: Routledge.

Wang, N. (1999) Rethinking authenticity in tourism experience. *Annals of Tourism Research* 26, 349–370.

Weigle, M. (1989) From desert to Disney World: The Santa Fe Railway and the Fred Harvey Company display the Indian Southwest. *Journal of Anthropological Research* 45 (1), 115–137.

Weigle, M. (1996) 'Insisted on authenticity': Harvey Indian Detours, 1925–1931. In M. Weigle and B.A. Babcock (eds) *The Great Southwest of the Fred Harvey Company and the Santa Fe Railroad* (pp. 47–59). Phoenix, AZ: The Heard Museum.

Weinstein, L. (ed.) (2012) *Native Peoples of the Southwest: Negotiating Land, Water, and Ethnicities*. Westport, CT: Praeger.

Whiteford, A.H., Peckham, S., Dillingham, R., Fox, N. and Kent, K.P. (1989) *I Am Here: Two Thousand Years of Southwest Indian Arts and Culture*. Santa Fe, NM: Museum of New Mexico.

Wright, M.N. (1989) *Hopi Silver: The History and Hallmarks of Hopi Silversmithing*. Albuquerque: University of New Mexico.

6 'Souvenirs' at the Margin? Place, Commodities, Transformations and the Symbolic in Buddha Sculptures from Luang Prabang, Laos

Russell Staiff and Robyn Bushell

It can be argued that objects of the everyday are only transformed into 'culture' when a semiotic process transforms quotidian objects into something else, a process of differentiation or 'marking out' often by 'outsiders' (cf MacCannell, 1976). The everyday textile production undertaken by Lao women in their homes, until relatively recently when clothing production was superseded by ready-mades from China, became 'Lao culture' when westerners regarded the aesthetic value of both the designs and the production of silk fabric as 'Art' to be displayed in museums (Bounyavong, 2001). However, this is a Eurocentric perspective and denies the way 'things' carry significance within the context of their production, distribution and ownership and often before any exchange associated with global market networks, particularly those created by, and dependent upon, tourism. The transformation of Lao weaving in the context of Western symbolic valuation and acquisition also alerts us to a conundrum: the use of the word 'souvenir' is problematic *vis-a-vis* material culture (are Lao textiles 'merely' souvenirs even when bought by tourists in a tourist setting?).

At the same time, souvenirs are marginalised within the western imaginary in a culturally produced hierarchy of material objects (art/design at the apex, kitsch/souvenirs at the bottom) and there is often a negativity accompanying the term because of a coupling with longing and nostalgia (Stewart, 1993). Indeed, the designation of any object as a 'souvenir' (objects related to remembrance and memory, a memento of travel) can be debilitating analytically even as it evokes powerful cultural and social resonances and critique (Hitchcock & Teague, 2000; Kwint *et al.*, 1999; Stewart, 1993). By examining the social life of carved Buddha sculptures, this chapter argues that the conception of the 'souvenir' is too limiting, and these objects, carry significance that is simultaneously central and marginal to the material culture of the heritage city of Luang Prabang, in Laos, because of the various transformations they undergo.

Our starting point is, therefore, not the souvenir per se (with its inescapable connotations relating to memory and remembrance) but material culture (and the social and cultural entanglements implied) more broadly conceived. Here, it is possible to discern a social/cultural *process* in Luang Prabang that produces souvenirs as multi-faceted objects that interrogate the commonly understood meaning of tourist keepsakes and in so doing can be simultaneously marginal and central. This argument requires two contextual overviews: Luang Prabang, the city (a place produced by local–global flows) and an ethnography of commodities.

Luang Prabang, Imagined and Represented

Luang Prabang is not just a city in northern Laos that is the physical habitat of its resident population and awaits the arrival of the traveller. Luang Prabang is the result of a series of mobilities and sedimentations; processes that create place from space and via the embodied interactions that integrate the physical and non-physical, the human and the non-human in a choreography of partially scripted and partially improvised forms (Baerenholdt *et al.*, 2004; Haldrap & Larsen, 2006, 2010).

For Lao people, Luang Prabang is imagined and represented on a number of simultaneous levels. It is first and foremost a sacred place, the physical incarnation of what it means to be a Buddhist in the Lao People's Democratic Republic (PDR), an historical and contemporary centre of Buddhist ritual and scholarship with its 34 monasteries, each one at the heart of a village (*ban*), and its population of up to 1000 monks (Heywood, 2006). Religious festivals and the mythology of the Naga, the protective water-serpent that is infused into the very landscape, contribute to the spiritual density of Luang Prabang (Ngaosrivathana & Ngaosrivathana, 2009; Trankell, 1999). The daily alms giving rite (*tak bat*) performed each morning at daybreak, creates a ritual geography of the city.

Mapped onto Luang Prabang as a religious centre are other imaginings. It is a former royal capital of a once extensive Lao-speaking kingdom (a history recently revived and recalibrated) (Evans, 2008; Trankell, 1999). It's the home of the *Pra Bang* Buddha, a talismanic image throughout Lao PDR and one of the foci of New Year celebrations, *Bun Pi Mai* (Stuart-Fox & Mixay, 2010).

Luang Prabang is a former French colonial town extensively rebuilt by the French in bricks and mortar in the first decades of the 20th century, now recast and re-presented in terms of the nationalist and revolutionary ideology of the current government (Grabowski, 2011; Long & Sweet, 2006; Pholsena, 2006). It is a market town and centre for higher learning with its university and colleges. It is a centre of health services, a river transport hub and, since inscription onto the World Heritage list (1995), a place of economic opportunity owing, in large measure, to the heritage-tourism nexus (Staiff & Bushell, 2013). These are some of the important ways Luang Prabang's dominant discursive coordinates represent the city to itself.

For the western visitor, the imaginings are in a different but connected register. Europeans have long represented South East Asia in ways deeply entwined with colonialism and 'Asia' has been 'orientalized' by administrators, travellers, scholars, writers and artists who created a European vision of 'the East' (Said, 1985). In the case of Luang Prabang, we have a continuum of writing beginning with the 19th century French explorer Henri Mouhot (1989) and today the travel and heritage descriptions of Heywood (2006) and numerous tourism websites.

In this imagining, Luang Prabang is enigmatic, seductive, lost in time, enchanting, unchanging, rooted in tradition, other worldly, spiritual and romantic (Staiff, 2012). What is of interest is the ways this western vision, initially through colonialism and now through representations of heritage/tourism, have indistinguishably criss-crossed with other imaginings that reticulate through the topography and reinforce an *idea* of Luang Prabang as a fusion of Lao and western (French colonial) characteristics. Put more bluntly, Luang Prabang cannot be imagined outside of the local and global dynamics that produce it. And, crucially, material culture cannot be conceptualised outside these social and cultural processes either.

An Ethnography of Commodities: 'The Social Life of Things'

In *The Social Life of Things* (1988), Arjun Appadurai outlines an ethnography of commodities. Following George Simmel's early 20th century work (and with a gesture to Karl Marx), Appadurai assumes that the value of an object is not innate but is a culturally and socially mediated judgement about value that is subjective, can be collectively sanctioned and is highly provisional (in that a value ascribed to things can change). For Appadurai (and Simmel), what extends 'thing-ness' to its life as a commodity is economic exchange because 'value is embedded in commodities that are exchanged' (Appadurai, 1988: 3). Commodities, therefore, exist in the space between desire/need and enjoyment/satisfaction and this space is bridged reciprocally in exchange by a process of 'exchange of sacrifice', what the parties are prepared to sacrifice to fulfil one's desire to exchange an object and to acquire an object.

The exchange process requires an infrastructure that enables mobility/ flows to translate an object from a 'local' thing to one that is globally redefined by its movement from one cultural and social setting to another. The translation is shaped by distance, difference, property rights, political regulation and re-contextualised semiosis, among other variables (Urry, 2007). Further, Appadurai argues that things can drop in and out of the commodity situation, that things may be 'differently characterised at different points in their social lives' (Appadurai, 1988: 13). However the conditions of exchangeability are critical. The exchange value is affected by how is it produced, who decides, what classification of objects is employed (is it 'art', 'antique', 'craft', 'textile', 'souvenir'?) and what knowledge systems underpin such classifications. Also, how the symbolic is produced and managed, how meaning and significance is produced and managed, and whether or not all parties in the exchange share values. This is especially the case when transactions occur across cultural boundaries. The politics of commodities arises from the regulation and management of the valuation and exchange processes. Identity issues arise because of the various ways exchange and its actors are considered socially with status, wealth and reputation being accrued by the selling of certain objects and the social capital accrued by the new owner. Appadurai introduces the notion of 'tournaments of value' (Appadurai, 1988: 21) as a way of indicating the importance of the performative aspects of exchange, the 'theatre' of the transaction, the ritual and the setting: bazaar, shop, auction, street-vendor.

Appadurai's (1988) ethnography of commodities applies to many cultural contexts and is grounded in particular historical and geographic circumstances and trajectories (pre/industrial/post-societies and/or western/non-western societies). The social life of a thing can be temporally, spatially and culturally multifarious, the same object having multiple lives or, 'a biography' of many identities (Appadurai, 1988). Importantly, traces of earlier lives can be retained. We explore this through the sculptured Buddhas of Luang Prabang by looking at symbolic retention, even under extreme conditions of transformation.

Today the west tends to regard objects as mute, inert, inanimate and thus objects are largely knowable in and through the processes (ritual, use, setting) and the symbolic order (including language) we attach to them. Appadurai (1988) points out this has not always been so. Despite capitalism and mass production, the pre-industrial conception of materiality still reverberates in the west even if curtailed, circumscribed or suppressed. Because of this resonance in occidental epistemology, there continues the capacity to understand, or at least appreciate, an ontology and teleology of object-hood that infuses material culture with a spiritual essence that cannot be divorced from an object's meaning, destiny, significance, power and use. This capacity is particularly important in the exchange of material culture under the aegis of western tourism in Luang Prabang because in Lao culture, spirits (*phi*) and spirit cults are ever present, and natural and supernatural phenomenon are fused indissolubly (Holt, 2009).

Commodities/Souvenirs: At the Intersection of Complex Social and Cultural Local/Global Processes

A multi-disciplinary study of material culture in Luang Prabang is potentially vast. In the context of local-tourism exchange of objects, we focus on just one example that intersects with our work (Staiff & Bushell, 2013): the carved wooden Buddha image. The production, distribution and sale of arts and crafts in Luang Prabang cannot be disentangled from a number of other cultural dynamics and social processes within which tourism and tourists are embedded. In this way, the object/commodity/souvenir is a highly visible symbol of a complex interplay of dynamically interwoven and sometimes contested phenomenon.

Carved wooden Buddhas for sale in Luang Prabang are predominantly in the style of two *mudras* (the iconographic positions of the Buddha), the 'Calling for Rain' posture (a standing Buddha with elegantly elongated arms to the side of the body, but the hands not touching the legs) and the 'Cease Fighting' posture (also a standing Buddha with both arms raised and extended with the palms of the hands facing the viewer) (see Figures 6.1 and 6.2).

The historical antecedents of these Buddhas can be found in Wat Visounnarath, originally built between 1512 and 1516 CE. The collection of early Buddha sculptures, presumably placed there after the temple was rebuilt in the late 19th century after a disastrous fire and looting of the city, is highly significant as it includes what may be 14th century Khmer influenced examples of the 'Calling for Rain' *mudra*. It is widely believed that Buddhism came under royal patronage in the 14th century (Stuart-Fox, 1998). The golden *Pra Bang* Buddha (in the 'Cease Fighting' *mudra*), after

Figure 6.1 Buddha images, 'Calling for Rain' *mudras*, wood, lacquer, gold leaf, 17th–18th century CE, Wat Visoun, Luang Prabang.

Figure 6.2 Buddha images, 'Cease Fighting' *mudras*, wood, lacquer, gold leaf, 16th–18th century CE, Wat Visoun, Luang Prabang.

which the city was named, is highly revered throughout Laos and was a gift from the Angkor royal family to King Fa Ngum (*c.* 1353–*c.* 1374 CE) (and currently viewed as a guardian spirit of Lao patriotism and independence) (Grabowski, 2011). Fa Ngum was married to the sister of the Angkor king, Phaya Sirichantra (Martin-Fox, 1998).

The 14th century wooden sculptures in Wat Visounnarath, therefore, carry high symbolic associations and along with the *Pra Bang* form a central part of the mythical and historical narrative of Laos nationhood, the royal past and Buddhism recently being re-inscribed into the imaginary of national identity by the Lao PDR Government (Evans, 2008, 1998; Pholsena, 2006). However, these sculptures carry more than historical value; they are talismanic. Positioned around the base of the principal seated image of the Buddha in Wat Visounnarath, they are part of a sculptural ensemble that is typical of all major Buddhist temples in the Theravada Buddhism tradition of Southeast Asia. These image ensembles are the carriers of singular concentrations of spiritual power at a place where the three axes of Buddhism come together: the Buddha, his teachings/the law, or *dhama,* and the community of monks, or *sangha* (Heywood, 2006; Holt, 2009). It is misleading, however, to only assert the distinctiveness of these two Buddha *mudras* in a place like Wat Visounnarath. Within the daily ritual of the monks, worship, festivals and the *wat* as a community centre for those who live in the village, of Visoun Narath, the sculptures are simply an on-going part of a complex sacred cosmology and practice that blend into the quotidian. In the daily lives of locals, they are not necessarily remarkable, historically or aesthetically (Holt, 2009).

While the age of these Khmer-influenced sculptures is not definite, the narrative associated with these early examples of 'Calling for Rain' is precise. The particular stylistic qualities and iconography of this *mudra*, as it developed in the royal capital of the Lao kingdom, is, according to locals, unique to Luang Prabang. Later examples of the 'Calling for Rain' Buddha in Wat Visounnarath, date from the 17th–18th century and bear witness to this place-based claim of uniqueness. Other historical examples of the *mudra* can be found in Wat Xien Thong, another monastery with strong royal connections, and until the French built a royal palace in the 20th century, one of the domiciles of the royal family when it was in Luang Prabang; and in the sacred caves of Pak Ou, overlooking the Mekong River upstream from Luang Prabang. Consequently, the 'Calling for Rain' and the 'Cease Fighting' Buddha images have contemporary significance within Luang Prabang and beyond, and metonymically, they connote Luang Prabang the city, its history, spirituality and identity. For Lao people, especially those living in Luang Prabang, these images of the Buddha are instantly recognisable and produce strong emotional affects/effects that are personal, religious and nationalist.

Claims of place-based uniqueness, however, have had a reach far beyond the country of Laos. Fuelled by earlier commentaries by French archaeologists, art connoisseurs, travellers and colonial administrators (Winter, 2007), along with western aesthetic tastes that, under the impact of Modernism, have applauded uniqueness, the 'Calling for Rain' Buddha and the 'Cease Fighting' Buddha effortlessly entered into occidental discourses of art, antiques and cultural heritage. Collectors quickly began seeking historical wooden Buddha sculptures like those to be seen at Wat Visounnarath and Wat Xien Thong. 'Uniqueness' had value across cultures but the ascribing of that value to the Buddha sculptures, during the exchange, was culturally specific on both sides of the ledger (and in a parallel and differentiated order of symbolic valuing and attachment).

At this juncture several things happened in the social life of the Luang Prabang Buddha images. The supply of Buddha sculptures considered to be 'historical' and that were in private hands (and thus available for exchange) was small and, according to unverified oral accounts, sculptures with an historical provenance were quickly bought and left the country. The interests of those engaged in the antique retailing business dovetailed, naturally, with the demand by western collectors, a demand that is complex and relates to the way Southeast Asian culture has been constructed and positioned in occidental representations. These include museum and art gallery exhibitions of Lao textiles and Buddha images, and within the design, media and film industries plus the on-going creation of the 'Asian exotic' within western tourism. However, these interests were circumscribed by the efforts of the Lao PDR Government to strictly curtail, through prohibition, the sale and export of 'heritage' Buddha sculptures and by the scarcity of such objects.

In the last ten years, owners of antique shops in Luang Prabang, usually foreigners (Chinese and Europeans), sometimes with Lao marriage partners or former Laotian citizens returning after voluntary exile post-1975, have responded to the western demand for the distinctive Luang Prabang *mudras* and in particular 'historical' Buddha images. They, together with the owners of luxury hotels in Luang Prabang, have commissioned local wood carvers to produce what, from a western perspective, may be called faked or copied antiques.

These highly skilled wood sculptors carve the Buddha in the two *mudras*, paint them (in a manner producing an incomplete look) and then bury the sculpture in a prepared pit of earth containing a special mixture of unspecified/secret ingredients. The sculptures are buried for one to two months. While buried, termites begin to attack the wood and the special mixture produces a patina of age. The product is a 'fake' historical Buddha, the historical value of the image marked by the greyish colour the wood becomes, the pitted surface, particularly the base of the sculpture – the outcome of termite activity – and the faded colour effect as though time has diminished the luster of the red and gold paint (see Figure 6.3). A few weeks produces a 'rare decade-old' antique!

Significantly, the Buddha sculptures are not blessed at a temple, a ceremony that mirrors the arrival of a new novice into the monastery. There are important reasons for this. In the home of the sculptors, women can handle the images. A woman cannot touch a blessed Buddha image, just as she cannot touch a monk. The sculptor can regard his commissions as 'merely work' with no spiritual overtones and obligations. The production is legal under Lao PDR law because they are not ritualised objects nor objects considered to be the patrimony of the nation. And, equally interestingly, the

Figure 6.3 Buddha images, sculptor's workshop, Luang Prabang, carved and aged in 2011. (The partially visible golden Buddha on the left is a private commission; the other two are for the antique shops in Luang Prabang.)

production of these Buddha images within Luang Prabang, and carrying symbolic associations tying them to the city, can be considered part of a wider political objective (see below).

Nevertheless, there is considerable ambiguity surrounding the status of these sculptures. Our research assistant and translator would not touch the Buddha images without showing a high degree of respect by performing a *wai* each time he came into physical contact with the sculptures. He later expressed concern that the Buddha images – one metre tall – were on the floor and therefore he was always towering above them, something he considered highly disrespectful. That the sculptures were not blessed did not, for him, lessen their spiritual import or the power of the image. He expressed the same concern about the clusters of wooden Buddha images in local antique shops. For many Laotians there cannot be any meaningful separation between the material object and its spiritual essence; every object has its associated spirit or *phi* (Holt, 2009), especially images of the Buddha.

The social life of the sculpture enters a critical phase in the context of exchange. In the retail shops the first transformation begins. The Buddha sculptures are, typically, arranged in clusters emulating the way the historical Buddha images are seen and photographed in Wat Xien Thong (a powerful and evocative image that has wide circulation on Luang Prabang websites, on travel websites, in guidebooks and in coffee-table books about Laos and replicated many times by tourists who visit this major 'attraction') (see Figure 6.4). This explicitly sets up a relationship between the historic Buddha sculptures and those newly made; historical authenticity through visual association.

For the buyers (and overwhelmingly the 'historic' Buddha images are for western collectors) a story of the sculpture's providence is told: the sculptures were hidden in caves and elsewhere during World War II to protect them and now they have been 'found'. (The story is not entirely a fabrication. According to locals in the retail business, the original found Buddha sculptures were hidden during the Japanese invasion, but they were small in number and the supply has long been depleted.) Consequently, buyers are told the sculptures for sale are dated 60 years old or older. So the 'exchange of sacrifice', that between the desire for a 'genuine antique' with an unambiguous symbolic connection to Luang Prabang (the two distinctive *mudras*) and the enjoyment of ownership, turns out to be about USD400+ for a one-metre tall wooden Buddha. This price is indicative. Some smaller Buddha sculptures sell for more than this, some taller ones for less. The critical value in this transaction is, first, the buyer's assessment of the signs of its historic value (the 'evidence' of age and acceptance of the World War II story), second, the distinctive poses of the figure that connect the sculptures with the discourse of Luang Prabang originality and the historical collections of the 'Calling for Rain' and

Figure 6.4 Buddha images, 'Calling for Rain' *mudras*, wood, lacquer, gold leaf, no date, Wat Xien Thong, Luang Prabang

'Cease Fighting' *mudras* and, third, an aesthetic assessment of the grace of the figure and the beauty of the face. (Interestingly, these aesthetic attributes of the Buddha are the same for both the Lao sculptor and the western buyer. And so there is an alignment between the Luang Prabang iconography, Lao aesthetics and western valuing.) For the seller anything above USD50, the price paid to the sculptor, is profit.

There is another dimension to this transaction. The Buddha sculptures of Luang Prabang carry more than the symbolic associations already discussed. An interesting question has always been, given the disquiet the Lao people of Luang Prabang have about Buddha images for sale to tourists as a commodity and the concern about the way they are displayed and handled in shops (an anxiety shared by the *sangha*, the community of monks), why has there been no attempt to regulate this trade? Recent scholarship about the Buddification of the Lao state by Evans (1998, 2008), Trankell (1999) and Pholsena (2006), suggest there is a national agenda here. The forging of a post-1975 national Lao identity has seen the revival of both royal and Buddhist commemorations and rituals in an attempt to create a national imaginary that is distinctive from Thailand (with its own substantial Lao-speaking Buddhist population in Issan or northeast Thailand) and conforms to the late 20th century mapped geographic identity of the country. However, for Pholsena, this project is fraught because only 75% of the Lao population is Buddhist (Pholsena, 2006: 71) and 75% of the population watch Thai television (Pholsena, 2006: 53).

Nevertheless, the conscious reformulation of the past and of Buddhism and the conscious appropriation of heritage into this nation-state discourse, means that the distinctive *mudras* of the Luang Prabang Buddha are also symbols of the Lao nation – the city being the original Lao kingdom. Indeed, the golden *Pra Bang* in the 'Cease Fighting' *mudras* explicitly carries this symbolism throughout Lao PDR. The production of the distinctive Luang Prabang Buddha sculptures for world distribution unconsciously projects for the buyer and consciously for the state, this imagining of the nation onto the world stage. And because of the economic benefits that accrue from the production and sale of Buddha sculptures, tourism is the vital vector in the exchange process.

The apotheosis of the transformation of the wooden Buddha sculptures, and another stage in the social life of the object, occurs when it enters the cultural world of its owner. We traced a pair of Buddha sculptures, one in the 'Calling for Rain' *mudra* and the other in the protection *mudra,* to Switzerland, where they are now displayed in the home of an art collecting couple (see Figure 6.5). The sculptures grace a sideboard in the entrance hall of the house. The metamorphosis into a purely western aesthetic object is complete in a place geographically and culturally remote from the site where the Buddha images were created and where they attached to themselves such powerful local symbolism. Of course, memories of Luang Prabang are adhered to the figures, in the classic way a souvenir is suppose to work, but the new context of display is far more telling. The sculptures exist in a house that displays contemporary Swiss art, European prints dating back as far as the 17th century, indigenous Australian paintings, wood carvings from Fiji, stone sculptures from Angkor, sculptures from Thailand, 19th century European portraits, landscapes and watercolours. It is a substantial private art collection. Travel is one axis running through this compendium but not the only one. In Appadurai's terminology, the commodity/souvenir phase of the life of the Luang Prabang Buddha images is now (temporarily) suspended. Most of the local references have been re-configured as curatorial narratives that give value to the sculptures as 'Art' in this globalised collection and provide a provenance in the family's electronic catalogue. The collectors are aware that the historicity of the sculptures may have been faked, but in the overall scheme of things this is of little concern. The attraction was the aesthetic appeal of the works and the desire to expand their art collection into Asian art. For the new owners, the value of the Luang Prabang Buddha images lies

Figure 6.5 Buddha image, 'Calling for Rain' *mudras*, wood and paint, purchased in Luang Prabang in 2009, private collection, Switzerland

in their relationship to the art collection overall, memory and remembrance being only a small part of the significance of the works newly displayed. Of particular importance is the way the iconography of the Buddha images is regarded. Deemed to be spiritual objects with a distinctive historical and symbolic narrative initiated by the *mudras* (and thereby absorbing the sculptures into western art history, theory and discourse), they are understood and appreciated as Buddha sculptures with sacred properties and Lao connections. However, this sense of the former symbolic life of the Buddha images is over-shadowed by the Swiss household decorative function the sculptures now have and the re-birthing of the sculptures as western aesthetic entities.

Conclusion: 'Souvenirs'? Margins or the Centre?

The Luang Prabang Buddha images in the distinctive *mudras*, reveal a complex entanglement of the local, global, communal, cultural, religious, economic and the social; an entanglement of place, tradition, history and modernity, of image and identity. Yes, they function, on one level, as souvenirs, a commodity exchanged in Luang Prabang with western visitors (Hitchcock & Teague, 2000; Kwint et al., 1999; Stewart, 1993). But the social life of these objects reveals that the concept of 'souvenir' is too limiting and unavoidably carries connotations that confine material culture to remembrance, memory, tourist keepsake and nostalgia. The Luang Prabang Buddha images in the 'Calling for Rain' and 'Cease Fighting' mudras are multidimensional objects: they are shaped pieces of wood, cut from the forest; they are symbols; at times, commodities and a mode of economic exchange; they are, in one 'chapter' of the biography of things, metonymic, having national and city-wide significance; they are an expression of 'living' cultural and social traditions; they hover, controversially, around the edges of the sacred; they are fakes; they are (western) art pieces; they are unavoidably entwined with tourism and the performance of tourists and so on.

Given that a margin is a relational concept, presuming 'a centre' to be meaningful, the Luang Prabang Buddha sculptures are interesting because, it can be argued, they simultaneously occupy both positions. For the wood carver, the object is profoundly local, the story of the sculpture beyond its creation having no impact on his and his family's lives. His work, within the economic order that transforms the Buddha images into commodities, displayed with a 'special' history in the main tourist street of the town, and to be sold to western travellers, is marginal. Indeed, they live on the very outskirts of Luang Prabang.

Within the World Heritage core of the city, the sculptures are a centre, linked, to various symbolic orders and discourses that give them a high degree of pre-eminence. However, in terms of global tourist flows and the global World Heritage system, Lao PDR is at the margins. Luang Prabang receives about 250,000 visitors a year. However, equally, the 'capture' of the Luang Prabang Buddha images by the global economy and by global flows, and by virtue of their relationship to the World Heritage system, places them at the centre of contemporary modernity and mobilities. Within the sign-ordered economy of western art, the sculptures are marginal and marginalised. Within the value-exchange system of the western art auction house, they are marginal. Their value as keepsakes is subjective and dependent on the way the sculptures are positioned by their owners; this could be central or marginal.

The question of margin/centre is, therefore, complex. The identity of the Luang Prabang Buddha sculptures as 'souvenirs' is limiting. What has been revealed is a series of entangled processes that are, perhaps, best understood as the 'social life of things', as material culture that is not and cannot be bound by the finality of typologies. The 'story' of the Luang Prabang sculptures is a narrative of constant change, fluidity, social and cultural dynamism and this is especially piquant for an object that began its life in a Buddhist place.

References

Appadurai, A. (ed.) (1988) *The Social Life of Things: Commodities in Cultural Perspective.* Cambridge: Cambridge University Press.

Baerenholdt, J., Haldrup, M., Larsen, J. and Urry, J. (2004) *Performing Tourist Places.* Aldershot and Burlington: Ashgate.

Bounyavong, D. (2001) Lao textiles: Past and present. In D. Kanlaya (ed.) *Legends in the Weaving* (pp. 8–27). Khon Kaen: The Japan Foundation Asia Centre.

Evans, G. (2008) Renewal of Buddhist royal family commemorative ritual in Laos. In F. Pine and J. de Pina-Cabral (eds) *On the Margins of Religion* (pp. 115–134). New York and Oxford: Berghahn Books.

Grabowski, V. (2011) Recent historiographical discourses in the Lao Democratic People's Republic. In V. Grabowski (ed.) *Southeast Asian Historiography: Unravelling the Myths* (pp. 52–69). Bangkok: River Books.

Haldrup, M. and Larsen, J. (2006) Material cultures of tourism. *Leisure Studies* 25 (3), 275–289.

Haldrup, M. and Larsen, J. (2010) *Tourism Performance and the Everyday: Consuming the Orient.* London and New York: Routledge.

Heywood, D. (2006) *Ancient Luang Prabang.* Bangkok: River Books.

Hitchcock, M. and Teague, K. (eds) (2000) *Souvenirs: The Material Culture of Tourism.* Aldershot and Burlington: Ashgate.

Holt, J. (2009) *Spirits of Place: Buddhism and Lao Religious Culture.* Honolulu: University of Hawai'i Press.

Kwint, M., Breward, C. and Aynsley, J. (eds) (1999) *Material Memories.* Oxford and New York: Berg.

Long, C. and Sweet, J. (2006) Globalization, nationalism and world heritage: Interpreting Luang Prabang. *South East Asia Research* 14 (3), 445–469.

MacCannell, D. (1976) *The Tourist: A New Theory of the Leisure Class* (reprinted 1999). Berkeley: University of California Press.

Mouhot, H. (1989) *Travels in Siam, Cambodia and Laos, 1858–1860* (first published in French in 1863). Singapore and New York: Oxford University Press.

Ngaosrivathana, M. and Ngaosrivathana, P. (2009) *The Enduring Sacred Landscape of the Naga.* Chiang Mai: Mekong Press.

Pholsena, V. (2006) *Post-war Laos: The Politics of Culture, History and Identity.* Chiang Mai: Silkworm Books.

Said, E. (1985) *Orientalism* (first published in 1978). London: Peregrine Books.

Staiff, R. (2012) The somatic and the aesthetic: Embodied heritage tourism experiences of Luang Prabang, Laos. In L. Smith, E. Waterton and S. Watson (eds) *The Cultural Moment in Tourism* (pp. 38–55). Abingdon and New York: Routledge.

Staiff, R. and Bushell, R. (2013) Mobility and modernity in Luang Prabang, Laos: Re-thinking heritage and tourism. *International Journal of Heritage Studies* 19 (1), 98–113. Available online DOI: 10:1080/13527258.2011.646287.

Stewart, S. (1993) *On Longing: Narratives of the Miniature, the Gigantic, the Souvenir, the Collection.* Durham and London: Duke University Press.

Stuart-Fox, M and Mixay, S. (2010) *Festivals of Laos*. Chiang Mai: Silkworm Books.

Stuart-Fox, M. (1998) *The Lao Kingdom of Lan Xang: Rise and Decline*. Bangkok: White Lotus Press.

Trankell, I. (1999) Royal Relics: Ritual and social memory in Louang Prabang. In G. Evans (ed.) *Laos Culture and Society* (pp. 191–213). Chiang Mai: Silkworm Books.

Urry, J. (2007) *Mobilities*. Cambridge: Polity Press.

Winter, T. (2007) *Post-Conflict Heritage, Postcolonial Tourism: Culture, Politics and Development at Angkor*. Routledge: Abingdon and New York.

7 Souvenirs as Transactions in Place and Identity: Perspectives from Aotearoa New Zealand

Jenny Cave and Dorina Buda

Souvenirs, as material cultures produced by hosts (and outsiders) for consumption by tourist 'others', are glocal expressions of place and identity, but have a range of meanings and authenticities for participants in the processes of production, sale and purchase. Our study suggests that retailers make choices about what to offer for sale to tourists based on their own interpretations of the region's geographic and cultural identity, as well as personal assessments of authenticities. In this chapter we theorise the position of suppliers in the context of local and global mobilities, and demonstrate how souvenirs can be interpreted in a post-structural frame. The chapter adds to the literature by providing a critical theory analysis of souveniring practices, meanings and institutions in the retail section of the tourism industry and offer a new understanding of how retail suppliers construct identities, market and sell souvenirs to tourists which Swanson and Timothy (2012) identify as a gap in literature. It also meets a gap in the literature on souvenirs depicting Aotearoa New Zealand.

Glocal Framing of Souvenirs and Retail

In a post-structural view of tourism, interactions between hosts and tourists are culture-bound and place-specific, affected by perceptions of 'self' versus 'other' and by the enterprise goals of the entrepreneur. Cave (2009a) theorised that access to art, material cultures and the natural resources that create cultural and place-specific identities embedded in two economies, the informal cultural and the formal market economy (Cave, 2009b). Further, interactions in tourism are framed as non-cash encounters in the informal economy and cash-based transactions in the formal market economy, influenced by the perceptions of cultural others and location-specific habitus of the host. Thus, interactions between cultural others span a range of utopic, heterotopic and dystopic experiences (after Foucault & Miskowiec, 1986).

Utopic experiences are idealised encounters between peoples of similar cultures to which access by outsiders (others) is privileged. They occur as idealised private spaces of togetherness where insiders associate, such as in cruise tourism, which is both critique of existing society and scripted ideal (Macbeth, 2000). Heterotopic experiences take place when foreign spaces are juxtaposed and superimposed, whether deliberately or unconsciously, to produce simultaneity of difference (Soja, 1995). They are quasi-eternal (museum or library), transitory (festival) or capable of suspending time (resort holiday), but are mediated by both the host and the responses of a guest. Dystopic experiences are the opposite of utopic and occur (or are avoided) in unfamiliar territory where people are strangers (to self). They can be radical, or conservative if a decline in values or loss of identity has occurred (Williams, 1988), such as in slum tourism (Dyson, 2012).

However, the enterprise domain in tourism is made up of values, internal structures and goals (after Bourdieu, 1991), encompassing the informal household economy in which identity and culture is maintained, as well as the formal market (Figure 7.1).

We situate our analysis in this framework of interactions (transactions and encounters), authenticity and commodification, as well as in the global–local relationship (Haldrup, 2009). We see souvenirs as glocal arte-facts of interaction between hosts (here suppliers of souvenirs) and guests. According to Giulianotti and Robertson (2007), the glocalisation process occurs in several ways depending on where people meet, within which social hierarchies and their values and receptiveness to others. The processes are relativisation, accommodation, hybridisation and transformation. The 'relativisation' of cultural practices occurs when prior expressions are retained, yet are differentiated from the new environment. 'Accommodation' maintains prior cultures, yet absorbs the practices of other societies. 'Hybridisation' produces distinctive hybrid forms that synthesise local and other cultural phenomena, whereas 'transformation' puts local culture aside in favour of alternative and/or hegemonic forms.

Furthermore, global travel activities exert a positive influence on tourists' beliefs about authenticity of crafts and craft product features (Yu & Littrell, 2003). On the one hand, experienced tourists have less tolerance

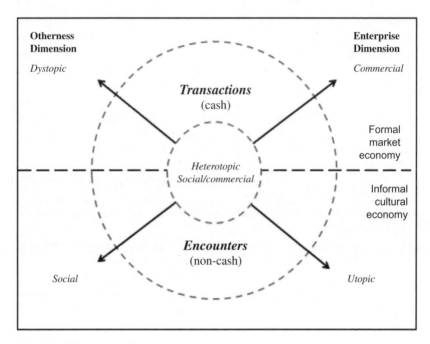

Figure 7.1 Theorising interactions in tourism enterprise
(After Cave, 2013)

with staged authenticity than those who have not travelled extensively (Littrell *et al.*, 1993). On the other hand, those tourists who are less experienced are most likely to understand authenticity as implying participation in the life of the local community and conform to the true/false, genuine/fake dichotomies (Waitt, 2000).

Consistent with our approach, post-structural theorists argue that authenticity is a chimera and reality is always mediated and simulated (Lane & Waitt, 2001; Waitt, 2000). Hence, material cultures produced within a culture in the utopic (self/same), social enterprise zone of Figure 7.1 are relativised, produced as authentic and consumed within cultural communities in market stalls and cultural festivals, but rarely in art galleries and souvenir shops, unless surplus to requirements (Cave, 2013).

However, the degree of authenticity or commodification of products and services in the heterotopic (self/other, othered same) social/commercial context is ambiguous and contested. Souvenirs of place and identity offered to tourists might use contemporary materials to render traditional forms, or traditional materials to produce contemporary ones; in glocalisation terms, representing accommodation and hybridisation. These are glocal forms, since they combine local attributes of place and identity with perceived global consumption requirements. They can be produced for sale but modified from the traditional original to preserve cultural knowledge and to enhance their appeal to tourist 'others'.

Thanatourism (Buda, 2012; Seaton, 1999) and high risk adventure tourism are located at the dystopic 'end' of the framework in Figure 7.1, paradoxically constructed as authentic and liminal experiences and commodified in souvenirs available for sale. Here, souvenirs might bear little resemblance to the authentic, but fully commercial, mass-produced, highly commodified and local traditional forms may be (mis)appropriated and transformed in favour of alternative and/or hegemonic forms such as global 'brands'.

Assumptions are made by retail suppliers of tourism experiences and souvenirs about the types of material culture that would 'interest' potential buyers. However, these can be flawed by inaccurate perceptions of the visitors' interest in representations of local place and culture (Cave, 2013). But do these theorisations hold for mainstream Anglo-European enterprise?

A Post-Structural View of Tourism and Souvenirs

The places that tourists visit are an integral part of how tourists construct, define and maintain their own identity, and say a great deal about their lifestyles, beliefs and the image that they wish to project to others (Williams, 2009).

Tourists collect physical artefacts (souvenirs) for themselves and for others as place-specific, local and multi-sensual 'touchstones of memory' (Morgan & Pritchard, 2005) that signify experiences and materialise self-identity, their choices reflecting perceptions of place (Gordon, 1986; Harrison, 2003; Swanson & Timothy, 2012); filtered however by the choices made by suppliers about what to make available to tourists. The word souvenir in French means 'to remember' and derives from the Latin verb 'subvenire', translating as 'occur to the mind' (Oxford English Dictionary, 2012). But as Swanson and Timothy (2012) note, this may be a person, a place, an occasion or an event, as well as a physical object. Thus, the souvenirs offered for sale are perhaps chosen to reflect specific significations that suppliers hope will prompt memories for the purchaser.

As tradable commodities, souvenirs are a major component of the retailing system in the formal marketplace, employing a significant number of people throughout the world in the production, distribution and sale. They can include items not intentionally produced as tourist mementos

(Swanson & Timothy, 2012) and contribute significantly to the economies of destinations (Timothy, 2005) since tourists spend about one-third of their total travel budget on shopping and buying souvenirs (Gordon, 1986; Harrison, 2003; Swanson & Timothy, 2012; Yu & Littrell, 2005). Souvenirs, in the main, are small, light-weight items (Fairhurst *et al.*, 2007; Kim & Littrell, 2001) that are commercialised replications of local customs (Cohen, 1988), ranging from primitive handicrafts to mass-manufactured items made in countries far from the destination where they are sold (Timothy, 2005). Further, gender and age are determinants of differences in purchasing behaviours. For instance, women in early adulthood frequently make unplanned purchases in malls while with children, while middle-aged women plan their shopping in speciality stores and tourist gift shops (Anderson & Littrell, 1995).

A number of researchers (Anderson & Littrell, 1995; Goss, 2004, 2005; Hashimoto & Telfer, 2007; Littrell *et al.*, 1993; Lury, 1997; Morgan & Pritchard, 2005) have investigated the symbolic and material roles of souvenirs. Gordon (1986) adopts a structural approach to the souvenir phenomenon in tourism, within binary oppositions such as home/away, ordinary/extraordinary and mundane/sacred, following Graburn (1977) and Cohen (1979). Gordon's (1986) typology of souvenirs: pictorial images, piece of the rock souvenirs, symbolic shorthand souvenirs, markers and local products, has been widely used. However the typology is rooted in a structural view of social science, in which cultural behaviour is characterised on binary or oppositional reference points, for which many examples do not fit exactly (Bird *et al.*, 1994). An alternative approach is to conceive of a range of classificatory reference points, around which meaning turns (Derrida, 1981), producing a perpetual duality of possibilities, a post-structural view.

Lury's (1997) critical analysis of souvenirs as objects of travel moves away from the binary opposition of home/away or dwelling/travelling to associate souvenirs with a hierarchy of travelers and dynamic global mobility, conception relevant to this study because of its potential application to a suppliers' perspective. Lury's traveler-objects, tripper-objects and tourist-objects are, in part, established by an authority of people's use of objects in their travelling and dwelling, and recognition of unique auras. 'Traveller-objects' sustain an authenticated connection to their place of dwelling and are characterised by their meaning and reference to their origin; examples include artwork, handicrafts and pieces of national political, historical or religious importance. In contrast, the significance and meaning of 'tripper-objects' is arbitrary, determined by the journey itself and the final dwelling place, such as, in a photo album, on a mantelpiece, on the refrigerator, etc. (becoming 'local' even if globally acquired). Mass-produced souvenirs, found objects and mementos are included as tripper-objects, as are shells and pebbles from a beach, tickets, matchboxes or personalised objects in the form of photographs and postcards. The third category, 'tourist-objects', include a wide range of objects from clothing, television programmes to food. These objects are constantly mediatised and mutually authenticating images and mobile objects 'in and of the in-between' (Lury, 1997: 80) because their purchasers/interpreters move within the practices of global cosmopolitanism. Global mobility thus generates mobility of objects, since objects often travel in conjunction with movements of people and so, in a way, cultures travel with and through souvenirs and are not a fixed set of objects, rooted in place (Urry, 2000).

'Object mobilities' cannot be considered fixed and performed in their unique meaning, but comprise a wide range of symbolic and material components and 'discrepant cosmopolitanism' reflective of diasporan rootedness in disparate locations, enduring changing global conditions (Clifford, 1994). Souveniring, for instance, plays an important role for Aotearoa New Zealand travellers during their OE, 'overseas experience' (Bell, 2002), and souvenirs are marketed more often as markers of place than of people (Asplet & Cooper, 2000). Yet very little analysis has been done in Aotearoa New Zealand with regard to souvenirs to date.

This study uses critical theory analysis to redress this gap asking: how do suppliers construct, market and sell souvenirs to travellers in the mainstream tourism industry?

Research Setting

The Bay of Plenty Te Moana a Toi on the east coast of the North Island of Aotearoa New Zealand was chosen for this study because of its relative isolation from tourism flows, the inseparability of Māori and European history from the natural landscape and the mix of domestic (local) and international (global) visitors. The bay was named by the seafarer Captain Cook because of its plentiful food, gentle climate and rich volcanic soils. It has had a history of intensive Maori settlement, flourishing around harbours, sheltered waterways, offshore islands and fortified hillside sites since 1500 AD, protected but isolated by the 600 metres high Kaimai-Mamaku Ranges. Although colonial and mission settlements were established in the 19th century and land wars were fought here, compared with the rest of Aotearoa New Zealand, European influence was late in arriving in the region (Cave & Law, 2008; Law, 2008), with the result that Māori culture remains strong and relatively undiluted.

Today, the rural area is home to 160,000 people, concentrated in the city of Tauranga, which provides deep water port services, as well as recreational marinas, boat moorings, yacht clubs and cruise ships (Shepheard, 2003). Tauranga has higher than the national average proportions of youth, retirees over 65 years in age and Māori, although ethnically Europeans dominate by two-thirds (Statistics New Zealand, 2012b). Geographically, the area also features beaches, thermal pools and extensive horticulture (kiwifruit, avocados) as well as the extinct volcano Mount Maunganui Mauao that overlooks a highly popular beach, apartment and shopping precinct of the same name, also home to the cruise ship terminal.

However, the Bay of Plenty Te Moana a Toi is peripheral to international and domestic tourism flows of the upper North Island. While in 2012, international guest nights in Aotearoa New Zealand totalled 6.5 million and domestic tourism 10 million, only 562,538 (3.4%) were spent in the Coastal Bay of Plenty (Statistics New Zealand, 2012a). The area is localised by topography, access routes and travel time (three hours by car from Auckland) and is outside the tourism 'golden triangle' of Auckland city, Rotorua (Māori culture) and the Waitomo Caves (underground caves). Visits to friends and relatives as well as holiday are the most important reason for international tourists to visit the area, many of whom are expatriate New Zealanders living in Australia and visitors from the United Kingdom. In contrast to the rest of Aotearoa New Zealand, where internationals outweigh domestic travellers by 22%, domestic tourism is the biggest market for the Bay of Plenty Te Moana a Toi, numbering 432,692 in 2012 (77%) (Statistics New Zealand, 2012a) arriving in the region from Auckland and the neighbouring Waikato for short breaks and holidays. In total, just under half a million people visit the area's beaches,and tourism attractions during the summer months of December to March, producing marked seasonality (Ministry of Tourism, 2011). Strategies to redress seasonality, increase and diversify international tourist numbers (Shepheard, 2003) are growing the cruise ship market from the Pacific Islands, North America and Europe, developing events with a more regional flavour to attract domestic and short haul (Australian) markets, utilising off-peak capacity and improving quality (Quality Tourism Development, 2010).

The region was designated as Coastal Bay of Plenty (CBOP) in 2006 to reflect the centrality of its local maritime characteristics in its tourism development strategy (Smart Tourism, 2006). The tourism products reflect this maritime setting and include heritage buildings, wharves, ferry and train excursions, tour operations, surfing, beach-based recreation, as well as museums, art galleries, events and activities, shopping malls, farmers' and craft markets, restaurants and accommodation. Using this case study area, the focus of this chapter is on the physical and symbolic meaning of souvenirs as a material culture that reflect maritime and land-based traditions, cultures and processes.

Critical Theory Analysis

This research adopts a critical turn to theorise place-associated meanings and identity as interdependent relations between local–global and socio-spatial processes such as mobilities (inwards and outwards from the destination) and products of socio-cultural constructions performed within a model of formal as well as informal transactions and encounters. Specific questions asked were: what are the meanings that sellers attach to souvenirs that they choose to offer for sale in terms of portrayal of place and identity? Does authenticity or commodification play a role?

In a critical, post-structural approach, perceptions of place and identity are thought of as social and cultural constructs. Writing descriptions and reflecting upon inter-subjective and material exchanges and social and non-human interactions (Watson & Till, 2010) helped us make sense of the ways that worlds are represented to the customer. Hashimoto and Telfer's (2007) inventory and observation process was used as a methodological base. Face-to-face structured interviews were completed with the managers of 41 souvenir retail shops in Bay of Plenty Te Moana a Toi. However interviewing is inextricably bound to its historical, political and geographic context and thus is not a neutral tool (Fontana & Frey, 2005). At one end of a continuum are structured interviews in which the interviewer asks lists of prescribed questions, while at the other end are unstructured interviews (Longhurst, 2009). Most of the questions that we asked were open-ended, asking if the global mobility customers (domestic tourists, locals or international tourists) made a difference to what was purchased and by whom, as well as establishing the meanings that attach to place and identity by how the suppliers' categorise the souvenirs they offer for sale. For example, inquiring whether any represent the region's Māori identity, specific geographies, local products grown or made in the Bay of Plenty, reflected its maritime character, or iconic 'Kiwiana'. 'Kiwiana' are popular cultural items that distinctly reference Aotearoa New Zealand (Bell, 2012) as a means to define identities of place and tourists. We also asked closed-ended questions about staffing, business type, owner/management structures and enterprise intent to establish where the businesses might be positioned in terms of the framework of tourist/host interactions in Figure 7.1.

Findings

Performed retailing

This study sampled all of the retail outlets in the CBOP, which advertised sales of gifts or souvenirs on the internet. The survey took place during the cruise ship season of November to February, to ensure that the work encompassed a period when suppliers had prepared stock for sale to local, domestic (glocal) and international (global) visitors.

Souvenir retailers in the western Bay of Plenty Te Moana a Toi are mainly small and medium businesses which include informal enterprise and registered retail business formats (one of which is indigenous). The range spans market stalls (micro-entrepreneurs and collectives) selling their own handicrafts and food as well as handicraft and fine art producers who operate small retail shop fronts. Businesses such as dealer galleries and retail stores in the formal tourism industry, gift shops in garden centres and retail shops that focus on art and/or gifts are represented in the sample. Publicly funded attractions and regional visitor destination information centres are also present.

All were staffed by at least one full-time person (the owner/operator), plus seasonal causal employees, but only the larger enterprises had waged personnel. Of the 41 locations, 19 (46%) were retail/gifts shops. Three had a strong focus on fine art, but the others varied from low-priced mass-produced items made offshore, to 'high-end art' by 'named' artists. Eight (20%) were producers who had developed a retail shop front from which to sell their work (weaving, pottery, photography, stained glass) or local produce (honey, kiwifruit) commodified into souvenirs targeted to tourist. Only one was an indigenous enterprise. Four were commercial dealer galleries that acted as agents for several local and national artists. Five were gift shops in attractions such as an art gallery, a private collection, museum or historic building; either fully or partially funded by public money, plus two regional visitor information centres. We also interviewed three art and farmer collectives in small gallery or café style outlets, which sell a smaller range of products and are operated by older artists and staff. The majority of souvenir retail outlets were located in Tauranga City (49%) and the beachfront suburb of Mount Maunganui (29%). The others were evenly distributed between smaller rural towns of Katikati (five shops), Te Puke (six shops) and the Kaimais (two shops) in the Coastal Bay of Plenty.

All enterprises hoped to attract international cruise tourists, but less than half had seen cruise ship visitors because of their constrained time and locations visited. Those located in the 'holiday spots' described a short period of global mobility in the form of international tourist visits during summer months, but a longer season of international visiting friends and relatives, some of who were expatriate New Zealanders living abroad and particularly United Kingdom relatives visiting new immigrants. Domestic tourists also were frequent customers and might be called glocally mobile, since many of them travel several hours to reach the Coastal Bay of Plenty. Many of the smaller outlets closed in winter and the larger ones shift their focus to provision of gifts and art to attracting local Bay of Plenty Te Moana a Toi residents to offset the seasonal downturn.

Portrayals of place and identity

Nine of the 41 stores created a thematic or colourful atmosphere that reflected either the place or the identity of the Bay of Plenty Te Moana a Toi. One retail/art store, for instance, used the koru (unfurling fern frond) symbol as a formal design element on the floor to define the display space and circulation of patrons. Another, a retail outlet for a horticultural producer specialising in the kiwifruit, organised the retail space to represent the oval-shaped fruit. Colour was the most frequently used wall, shelf and flooring device to separate types of goods or to identify staff and highlight signage. For example a honey producer used yellow, brown and black for display purposes, and a Māori retailer used 'found' objects from the beach and bush as a backdrop for indigenous goods. Most display areas, however, were treated as functional, shelved spaces that relied on colourful stock, arranged in product groupings or discount messaging to attract visitors into the store. There was tendency towards clutter and difficulty in separating out and locating items.

Thus little effort was made by the retailers to interpret the cultural or physical landscape to the customer. None of the goods, or display spaces was themed, nor was colour used to reflect the deep maritime or Māori cultural heritage of the region. The suppliers did not take advantage of the local natural scenery outside the shop (beaches, bush and water) as display surroundings. There is a gap then in the tourism retail market of the Bay of Plenty Te Moana a Toi for themed stores and for an 'all Bay of Plenty-made' souvenir shop.

Meanings attached to place and identity

We interviewed suppliers about their categorisation of the items offered for sale as icons of the region, cultural identity or natural environment. We also observed the range of stock provided on the shelves and inventoried the items offered for sale to provide examples of these items. The results of the interviews and observations are contained in Table 7.1, which shows business

Table 7.1 Types of enterprise, supplier interpretations and object mobilities

Theme / Retail type	Bay of Plenty	Māori	Maritime	'Kiwiana' — Traveller objects
'Farmer'/art collective	Vegetables. Meats. Honey, pickles, jam. Breads.	Organic produce. Maori potatoes. Wood carvings.	Smoked shellfish & fish.	Woven and knitted wool. Clothes. Beading. Carved wood. Candles.
Producer/ retailer	Honey products. Kiwifruit products. Avocado cosmetics. Kiwi 360 (liqueur, wine).	Wood carvings. Traditional clothing. Woven baskets. Maori (tiki, koru, waka, haka) symbols in modern materials.	Seaside & coastal photos. Stained glass.	Handmade paper. Woven & knitted wool. Wood carving. Flowers photos. Metal sculptures. Iconic symbols (tiki, kiwi, koru).
Garden centres	Art products made locally.	Koru & silver fern artwork. Māori motif garden sculptures. Key rings. Fridge magnets.	Coastal plants. Beach bags & hats. Jewelry. Maritime theme plates. Shell design fabric.	Pukeko ornaments. Tin art. Household gifts. Sculpture. Native seeds. Toiletries.
Attraction	Local postcards. Historical books.	Māori culture books.	Ceramic tiles.	Scenic postcards.

(continued)

Table 7.1 (*Continued*)

Retail type	Bay of Plenty	Māori	Maritime	'Kiwiana'
	Tripper objects		Tourist objects	
Visitor information centre	Postcards. Tea towel. Kiwifruit soap & chocolates. Thermal mud cosmetics. Mount Maunganui t-shirt. Food products.	Carved jewelry (bone, wood, pounamu & shell). Plastic Māori artwork. Weaving & carving.	Pāua shells, Beach pebbles. Postcards. Posters.	Postcards. Confectionary. Clothing/t-shirts. Kiwi soft toys. Kiwiana icons in wood, glass & ceramics.
Dealer gallery	Photos of Mount Maunganui. Historic photos of daily life.	Carvings.	Seascape and fishing photos. Fish paintings, oils & watercolour.	Churches, heritage buildings. Painted and mixed media art.
Retail/art	Photo blocks of seascapes, coastal flowers. Posters of Mount Maunganui. Paintings of beach, seascapes.	Wooden carvings. Woven baskets. Ceramics. Metalwork. Painted and mixed media art.	Pāua jewelry. Bone carvings. Paintings of beach, seascapes.	Oamaru stone sculptures. Merino/opossum clothing. Shell, pounamu and silver jewelry. Ceramic tiles with kiwiana images.
Retail/gift	Kiwifruit soap & chocolates. Thermal mud cosmetics. Mount Maunganui t-shirt.	Carved jewellery (bone, wood, pounamu and shell). Māori culture-themed wall hangings and artwork.	Pāua shells. Postcards. Posters.	Postcard, tea towels. Confectionary. Clothing/t-shirts. Cosmetics. Pukeko, kiwi, 'Buzzy Bee', 'All Blacks' brand. Pohutukawa flower symbols on soft toys, wood, glass & ceramics.

type by retail suppliers' categorisations, as well as indicates representations of traveller (authenticated connection), tripper (connected arbitrarily by virtue of the journey and final dwelling place) and tourist objects (mutually authenticating objects constantly connected to glocal processes) (per Lury, 1997). The separation into categories is not perfect and there is some overlap.

As shown in the table, we often found that souvenirs categorised by managers as 'maritime' overlapped with the Māori culture category, such as pāua shell (abalone) souvenirs, whether in forms of shells, inlaid in wood-carvings and as jewellery. The research found that not only tourists shop for pāua objects, but locals do as well. The shell is not only relegated to ashtrays and tacky jewellery for sale to tourists (Light, 2003), but decorates fashion accessories, shoes, belts and jewellery. However managers from two shops regarded pāua objects as significant to Māori culture. The types of pāua present in the Bay of Plenty Te Moana a Toi area are from the family of Haliotis iris, but the snail is black and the shell has strong, vibrant and iridescent colours. Māori people regard pāua as a taonga (treasure) because it is gifted to them by Tangaroa, the God of the Sea. In relation to Gordon's typology, pāua souvenirs 'cross the boundaries' of Gordon's souvenir categorisation by combining natural and manufactured qualities. They are not fully 'piece-of-the-rock' but also a 'symbolic shorthand' that evokes a coded or shorthand message about place or time (Gordon, 1986).

Some well recognised symbols of Aotearoa New Zealand culture are pāua, kiwis (long-beaked flightless birds), silver ferns (a silvered tree fern), korus (a spiral shape based on the unfurling silver fern frond), pounamu (a hard nephrite rock also called 'greenstone') and red socks (worn to support the Aotearoa New Zealand Americas Cup yachting campaign). As Light (2003) says 'other icons don't capture the crisp clarity of days when the southerly blows or the beauty of our long and varied landscapes' (Light, 2003: 72).

Many of the kiwiana souvenirs were commodified versions of iconic items, but labelled as Aotearoa New Zealand made or made from iconic local materials. For example:

A number of companies producing goods such as t-shirts state on the label that they are designed, marketed in NZ and are made in China, some say 'All NZ made' (but you can feel a different product quality). A lot are more expensive. For example, for one paua shell range of jewellery (pendants, necklaces, bracelets, and earrings) the paua shell is collected in NZ, shipped to China and made into the jewellery and sent back to NZ to be sold. This keeps the price to between $15–20. Whereas a product that is all made in NZ sells from $20–25. This means that the foreign made product will sell more cause simply of its price. (Art gallery manager)

The majority of symbolic souvenirs commercialised in these shops inhabit the same in-between space and assume more than one identity trait. The haka (Māori ritual challenge) appears on t-shirts, towels and pens as words, and warrior images, carved bone jewellery and pounamu mimic prehistoric carved whalebone and nephrite forms, Oamaru stone (limestone the South Island of Aotearoa New Zealand), glass work and pottery carved into the koru form, mixed media and painted artwork and soap made with kiwifruit (edible berry sold commercially under the 'kiwifruit' name), all cross those boundaries between category.

We argue that such souvenirs perform more than a marker's function or symbolic shorthand and so are traveller-objects identified by Lury (1997) that travel well, maintaining their meaning across time and space, across different localities within an increasingly globalised world. Pounamu and koru, for example, retain a glocal identity as they keep reference to their original Aotearoa New Zealand origin and form. The word itself, has had a copyright placed on it so that only green nephrite may be called pounamu (Te Kaha, 2012).

Postcards, posters, paintings and photos of flowers, the seaside, the maritime coast or fish, as well as prints with shells, represent another significant part of the souvenirs bought by tourists in the Coastal Bay of Plenty. Other examples are images 'Buzzy Bee' (striped pull-toy), 'All Blacks' (Aotearoa New Zealand's rugby teamf), pohutukawa flowers (red stamen) and pukeko (black, red and white flightless bird), on soft toys, wood, glass and ceramics. These are called 'pictorial images' types of souvenirs by Gordon (1986) and tripper-objects by (Lury, 1997). The postcard and the photograph in our survey were among the top five best-selling souvenirs at all of the shops under analysis. The meaning of such souvenirs is constantly reconstituted and negotiated by their final destination, their resting place, whether on a mantelpiece or in an album and so carries on open, imposed from outside the object by external context (Lury, 1997).

Another category is that of local products in Gordon's (1986) vocabulary or tourist-objects in Lury's (1997). Honey, skincare products, wood products, sheepskin products, Māori potatoes introduced to Aotearoa New Zealand by seafarers in the late 1700s (Leach, 1984), vanilla extracts, avocado oil and macadamia nuts are some examples of the local products consumed as souvenirs. The movement and meaning of these tourist-objects are different than those of tripper-objects discussed above, as their traits are neither open nor closed 'but in-between, there and here in their journeying' (Lury, 1997: 79).

Meanings attached to souvenirs

When asked why they sold these items as souvenirs, the majority of managers said that they wanted to gather together local products as souvenirs for sale to tourists and gifts for local consumers. Local souvenirs were objects of maritime and non-maritime nature, typically representing the food and natural products of the Coastal Bay of Plenty region and/or Māori culture.

We try to sell a number of items and of these ones that are NZ made as we have found the customers demand NZ made. However, the customers also demand a low price, of which NZ-made can't compete on. (Attractions manager)

The producers/retailers are very focussed on authenticity and targeted an upper range of priced products. Five of the organisations interviewed were art/craft collectives. Some operated in small towns in the local area, but the largest was in the centre of the city, targeting cruise tourists, but open on restricted hours.

The findings of this project suggest that tourist identities performed in the Coastal Bay of Plenty are also a mixture of fascination and interest with maritime and Māori values, expressed in a fine weave of coastal and Māori values that reflects in the commercialisation of souvenirs. However, such values constitute a larger national image that does not specifically differentiate this particular destination from others in Aotearoa New Zealand. This has implications not only for the perceived 'authenticity' of experience at the destination, but also for supplier competiveness and long-term survival.

The development of gift retailing targeted to the international (global) tourist, expatriate New Zealanders and national visiting friends and relatives (glocal) and residents (local), may be one way to offset deficits of new money in the local economy, geographic peripherality and seasonal reliance on domestic tourism and family visitors.

Further, the findings noted in Table 7.1 suggest that representatives of the retail industry in the CBOP who are most closely linked to the informal producer economy, the farmer/art collectives and producer/retailers, do supply souvenirs that take advantage of the region's extreme maritime beauty, local artists and Māori identity.

However, those that are more commercial, such as the retail/gift shops, in their outlook, rely instead on supplying generic material culture (souvenirs) that are iconic to the nation of Aotearoa New Zealand, but do not effectively differentiate the destination and paradoxically produce price competitiveness. Yet, this is at odds with the evident demand from glocal tourist and local consumers for place and culture-specific items that reinforce the experience in that location. The suppliers, who locate in between these, are perhaps more hybrid in intent and emphasise authenticity of culture and direct connection to place, offer a range of all types of souvenir from postcards to products made by locals, in the local area, as well as the generic 'kiwiana' items that are not differentiated.

Overall, we found that over half of the organisations did stock Aotearoa New Zealand made/produced products, but around one quarter specialised in products that were made in the Coastal Bay of Plenty area which is a small niche market that is growing slowly. There are a number of products that are labelled as if they are designed and marketed by Aotearoa New Zealand people and organisations, but many are made in China. There are large numbers of products made in China to look like Kiwiana souvenirs, and some that use local materials, such as pāua, are produced cheaply in China and then sent back to Aotearoa New Zealand, making it difficult for other companies to compete as it affects price, number of products made/produced and quality with local products mainly being of higher quality and material quality. These objects associate with a hierarchy of tourists and a dynamic of global mobility between local and international places, intermediated however by visiting friends and relatives, who are expatriate New Zealanders from Australia.

Conclusion

This chapter contributes to our understanding of place-associated meanings and formation of glocal destination identity. It theorises meanings of authenticity and commodification, place and identity as sociological and geographic interpretations of the interdependent relation between local–global and socio-spatial processes. The research contributes an operationalisation of the significance of material cultures, such as souvenirs, in the formal tourism industry, specifically in the retail sector. The study also goes some way towards addressing a gap in the literature about how the ways that retailers construct identity and choose products project their identity to consumers, as well as the role of retail in the tourism experience. Further it adds to the literature on the depiction of national versus local identity to global and glocal audiences and expectations of the impact of the cruise industry.

This chapter contributes original thinking to the literature on tourism souvenirs in five ways. First, by conceptualising a post-structural framing of souvenirs within tourism as glocal (mis)appropriations of authenticity, place, heritage, identity and brand, and second, by adding to understandings of the social, cultural and economic construction of souvenirs that remains largely underexplored. Third, the chapter located these concepts in a range of formal and informal retail experiences. Fourth, it situates the consumption of authenticities to glocal mobilities. Fifth, it is a unique case study in a region of Aotearoa New Zealand where Māori culture and imagery is intertwined with tourist consumption. This has implications for further research to untangle the complexities of identity and representation of this country where ethnically, Māori are 12% of the population, alongside growing numbers of Pacific Island and Asian people and declining numbers of people of Anglo–European descent.

At an industry level, these findings have implications, not only for the authenticity of experience at the destination where the study was undertaken, but also for retail supplier competiveness and their long-term survival in the tourism marketplace. Finally, the research sets a baseline of hypothesised relationships, against which a further study will assess consumers' response to the goods made available.

References

Anderson, L. and Littrell, M.A. (1995) Souvenir purchase behaviour of women tourists. *Annals of Tourism Research* 22 (2), 328–348.

Asplet, M. and Cooper, M. (2000) Cultural design in New Zealand souvenir clothing: The question of authenticity. *Tourism Management* 21 (3), 307–312.

Bell, C. (2002) The big 'OE': Young New Zealand travellers as secular pilgrims. *Tourist Studies* 2 (2), 143–158.

Bell, C. (2012) Kiwiana goes upmarket: Vernacular mobilization in the new century. *Continuum: Journal of Media & Cultural Studies* 26 (2), 275–288.

Bird, J., Curtis, B., Mash, M., Putnam, M., Robertson, G. and Tickner, L. (eds) (1994) *Travellers Tales: Narratives of Home and Displacement*. London: Routledge.

Bourdieu, P. (1991) *Language and Symbolic Power* (G. Raymond & M. Adamson, Trans.). Cambridge, MA: Harvard University Press.

Buda, D.M. (2012) *Danger-zone Tourism: Emotional Performances in Jordan and Palestine*. PhD, Waikato University, Hamilton, NZ.

Cave, J. (2009a) *Between Worldviews: Nascent Pacific Tourism Enterprise in New Zealand*. PhD. University of Waikato, Hamilton.

Cave, J. (2009b) Embedded identity: Pacific Islanders, cultural economies and migrant tourism product. *Tourism, Culture and Communication* 9, 65–77.

Cave, J. (2013) Pacific migrant cultural enterprise: Challenges and solutions. *Arbeitshefte Quaderin* 46, 103–116.

Cave, J. and Law, G. (2008) Appendix 5. Cultural tourism and archaeology in the Bay of Plenty Conservancy. In G. Law (ed.) *Archaeology of the Bay of Plenty* (pp. 141–143). Wellington, New Zealand: Department of Conservation.

Clifford, J. (1994) Diasporas. *Cultural Anthropology* 9 (3), 302–338.

Cohen, E. (1979) A phenomenology of tourist experiences. *Sociology* 13, 179–201.

Cohen, E. (1988) Authenticity and commoditization in tourism. *Annals of Tourism Research* 15, 371–386.

Derrida, J. (1981) *Positions*. Chicago: University of Chicago Press.

Dyson, P. (2012) Slum tourism: Representing and interpreting 'reality' in Dharavi, Mumbai. *Tourism Geographies* 14 (2), 254–274.

Fairhurst, A., Costello, C. and Holmes, A. (2007) An examination of shopping behavior of visitors to Tennessee according to tourists typologies. *Journal of Vacation Marketing* 13 (4), 311–320.

Fontana, A. and Frey, J.H. (2005) The interview: From neutral stance to political involvement. In N.K. Denzin and Y.S. Lincoln (eds) *The SAGE handbook of qualitative research* (pp. 695–728). Thousand Oaks, CA: Sage.

Foucault, M. and Miskowiec, J. (1986) Of other spaces. *Diacritics* 16 (1), 22–27.

Giulianotti, R. and Robertson, R. (2007) Forms of glocalization: Globalization and the migration strategies of Scottish football fans in North America. *Sociology* 41 (1), 133–152.

Gordon, B. (1986) The souvenir: Messenger of the extraordinary. *Journal of Popular Culture* 20, 135–146.

Goss, J. (2004) The souvenir: Conceptualizing the object(s) of tourist consumption. In A.A. Lew, C.M. Hall and A.M. Williams (eds) *A Companion to Tourism* (pp. xviii, 622). Malden, Mass: Blackwell.

Goss, J. (2005) The souvenir and sacrifice in the tourism mode of consumption. In L. Cartier and A.A. Lew (eds) *Seductions of Place: Geographical Perspectives on Globalization and Touristed Landscapes* (pp. 56–71). London, UK: Routledge.

Graburn, N. (1977) Tourism: The sacred journey. In V. Smith (ed.) *Hosts and Guests: The Anthropology of Tourism* (pp. 17–31). Philadelphia, Pennsylvania: University of Pennsylvania Press.

Haldrup, M. (2009) Local–Global. In K. Rob and T. Nigel (eds) *International Encyclopedia of Human Geography* (pp. 245–255). Oxford, UK: Elsevier.

Harrison, J.D. (2003) *Being a Tourist: Finding Meaning in Pleasure Travel*. Vancouver: University of British Columbia Press.

Hashimoto, A. and Telfer, D.J. (2007) Geographical representations embedded within souvenirs in Niagara: The case of geographically displaced authenticity. *Tourism Geographies* 9 (2), 191–217.

Kim, S. and Littrell, M.A. (2001) Souvenir buying intentions for self versus others. *Annals of Tourism Research* 28 (3), 638–657

Lane, R. and Waitt, G. (2001) Authenticity in tourism and native title: Place, time and spatial politics in the East Kimberley. *Social & Cultural Geography* 2 (4), 381–405.

Law, G. (2008) *Archaeology in the Bay of Plenty*. Wellington, New Zealand: Department of Conservation.

Leach, H.M. (1984) *1,000 Years of Gardening in New Zealand*. Wellington: Reed.

Light, E. (2003) Paua to the people. *North & South* 203, 72.

Littrell, M.A., Anderson, L.F. and Brown, P.J. (1993) What makes a craft souvenir authentic? *Annals of Tourism Research* 20 (1), 197–215.

Longhurst, R. (2009) Interviews: In-depth, semi-structured. In R. Kitchin and N. Thrift (eds) *International Encyclopedia of Human Geography* (pp. 580–584). Oxford: Elsevier.

Lury, C. (1997) The objects of travel. In C. Rojek and J. Urry (eds) *Touring Cultures: Transformations of Travel and Theory* (pp. 75–95). London: Routledge.

Macbeth, J. (2000) Utopian tourists – Cruising is not just sailing. *Current Issues in Tourism* 3 (1), 20–34.

Ministry of Tourism (2011) Regional tourism data 2005–2010. Retrieved from: http://www.tourismresearch.govt.nz/By-Region/Regional-Data.

Morgan, N. and Pritchard, A. (2005) On souvenirs and metonymy: Narratives of memory, metaphor and materiality. *Tourist Studies* 5 (1), 29–53.

Oxford English Dictionary (2012) 'souvenir, n.' Retrieved from http://www.oed.com/.

Quality Tourism Development (2010) Bay of Plenty Tourism Performance and Future Opportunities Report (Draft 3.0).

Seaton, A.V. (1999) War and thanatourism: Waterloo 1815–1914. *Annals of Tourism Research* 26 (1), 130–158.

Shepheard, N. (2003) Smart Port the Tauranga way. *North & South* 205, 66.

Smart Tourism. (2006) *Bay of Plenty Tourism Strategy*. Tauranga, New Zealand: Tauranga City Council.

Soja, E. (1995) Heterotopologies: A remembrance of other spaces in the Citadel-LA. In S. Watson and K. Gibson (eds) *Postmodern City Spaces* (pp. 14–34). Oxford: Blackwell.

Statistics New Zealand (2012a) *International Travel and Migration: September 2012*. Wellington, New Zealand: Statistics New Zealand.

Statistics New Zealand (2012b) *QuickStats About Western Bay of Plenty District*. Wellington, NZ: Statistics New Zealand.

Swanson, K.K. and Timothy, D.J. (2012) Souvenirs: Icons of meaning, commercialization and commoditization. *Tourism Management* 33, 489–499.

Te Kaha. (2012) Te Kaha Pounamu. Retrieved from http://www.tekahapounamu.com.

Timothy, D.J. (2005) *Shopping Tourism, Retailing and Leisure*. Clevedon: Channel View Publications.

Urry, J. (2000) Sociology Beyond Societies: Mobilities for the Twenty-first Century London: Routledge.

Waitt, G. (2000) Consuming heritage: Perceived historical authenticity. *Annals of Tourism Research* 27 (4), 835–862.

Watson, A. and Till, K.E. (2010) Ethnography and participant observation. In D. De Lyser (ed.) *The SAGE Handbook of Qualitative Geography* (pp. 121–137). Los Angeles, CA: Sage.

Williams, D. (1988) Ideology as dystopia: An interpretation of "Blade Runner". *International Political Science Review* 9 (4), 381–394.

Williams, S. (2009) *Tourism Geography: A New Synthesis*. London: Routledge.

Yu, H. and Littrell, M.A. (2003) Product and process orientations to tourism shopping. *Journal of Travel Research* 42 (2), 140–150.

Yu, H. and Littrell, M.A. (2005) Tourists' shopping orientations for handicrafts: What are key influences? *Journal of Travel and Tourism Marketing* 18 (4), 1–19.

Part 3

Glocal Case Studies in Sustainable Tourism

8   Green Tourism Souvenirs in Rural Japan: Challenges and Opportunities

Atsuko Hashimoto and David J. Telfer

In rural Japan, government-sponsored Green Tourism projects are being developed as a way to rejuvenate aging and declining agricultural-based economies. Green Tourism represents holiday activities engaged in while staying at a farming village, a fishing village and/or a forestry village according to the Japanese Ministry of Agriculture, Forestry and Fisheries. On the Kunisaki Peninsula, in the Prefecture of Oita on the island of Kyushu, Green Tourism projects range from agritourism-based hotels, farmers' markets, agricultural eco-museums, bed and breakfasts, and hot-spring spas. This chapter will examine the efforts of local woman residents to create innovative souvenirs to be sold at a variety of shops and attractions throughout the region. It is based on on-going field research since 2003. The chapter will not only look at the products sold, but also the challenges faced by the souvenir producers and the relationship to female empowerment. The souvenirs produced by local farmers include food products as well as craft items, many of which are made from recycled materials such as buckwheat pillows. One group of woman farmers is preparing authentic traditional Japanese lunch boxes (*bento*) using local ingredients that are to be sold at a nearby area heritage tourism district. Some of the challenges faced by these aging rural farmers producing the souvenirs include time conflicts with traditional farming, seasonality of tourism and lack of support from family members.

Souvenirs and Female Empowerment

Souvenirs have multifaceted roles of depicting images, reinforcing identities, providing employment, triggering positive memories of people's holidays and yet they can also be linked with the commodification process associated with souvenirs and handicrafts (Swanson & Timothy, 2012). The complexity of the nature of the relationship between souvenirs and local culture can be further viewed in terms of locals demonstrating and reviving elements of their culture and traditions, or even as creating new forms of tourism souvenirs (Markwick, 2001). Clearly souvenirs are much more than objects that often sit on a shelf collecting dust weeks or months after the holiday is over. They have important implications for those that are producing and selling the souvenirs to tourists and can influence the resulting power relationships both inside and outside the local community. As a new source of income, souvenirs can potentially represent important sources of funds and instil a sense of self-reliance and empowerment, yet this may be too simplistic of a view. In the case of textile production in Oaxaca, Mexico, Cohen (2001) found that local souvenir-based projects tended to favour certain classes of producers while impoverishing others, and therefore limiting tourism as an agent of community development. In the context of indigenous communities dangers of cultural appropriation (Blundell, 1993) further complicates the relationship between tourist souvenirs and local communities.

Within the context of this chapter, the authors are investigating, in part, the changes in the role of women in rural Japan as a result of producing souvenirs for tourists. There is a small but growing number of studies that link souvenir production and women's empowerment (e.g. Apostolopoulos et al., 2001; Davis, 2007; Hitchcock & Kerlogue, 2010) and this study will add to this body of literature. At a broader level, Ferguson (2011) also points out, there has been little work conducted on the ways in which tourism can contribute to the third UN Millennium Development Goal, which is to promote gender equality and empower women. Hashimoto (2002) argues the empowerment of women through tourism can be examined as an agent of change for family structures, as well as in the context of power structures in society. In some traditional societies women may not have been the main breadwinners and when they start earning a higher salary from selling souvenirs through tourism they may earn more than men in the primary industries (i.e. agriculture or fishing), possibly creating conflicts (Hashimoto, 2002; Hashimoto & Telfer, 2011). In examining the changing lives of Mayan craftswomen, Cone (1995) illustrates how the women responded to the opportunities from tourism, which transformed their traditional patron–client role to a modern form of expressive friendship. The two craftswomen in her study rejected the traditional Mayan women's role for themselves and they had 'constructed their lives, their identities, their life stories, with a sense of adventure and of increasing confidence in their abilities to rise to new occasions and encounters' (Cone, 1995: 325). Cone (1995) also found that the type of product is also important to consider, as Mayan woman potters had considerably more commercial experience than the woman weavers, having participated in marketing their craft.

The importance of souvenir sales by women is reflected in the study on Tonga by Connelly-Kirch (1982) who found that 80% of those selling souvenirs to the passengers from cruise ships were women and this had been partly organised through a cooperative. Similar findings were discovered in Micronesia (Nason, 1984), that as the handicraft market and its value increased with the expansion of tourism, more and more women became the primary producers of handicrafts.

The challenges of performing multiple tasks for women souvenir producers was also identified by Connelly-Kirch (1982), as most of those interviewed had to sandwich handicraft production between other daily responsibilities, including housecleaning, cooking, child care and participation in village cooperative women's activities (Connelly-Kirch, 1982). This notion of women having a double duty of working in the tourism industry as well as being expected to maintain the domestic household has also been found by Hashimoto and Telfer (2011) and Costa (2005).

Swain (1993) provides a critical comment on gender and the production of souvenirs and makes an important observation based on women producers of ethnic arts in Latin America and China. She argues that gender dynamics of indigenous ethnic arts production cannot be explained by development theories. Development theories predict that either women will be empowered by economic gain or will be exploited by the patriarchal drive of global capitalism, which is epitomised by international tourism. Rather, Swain found that internal (family/community) factors enabled women's empowerment while the external factors (market/state) in a stratified gender/class/ethnicity order, limit role options for indigenous women and men. Clearly the relationship between women and souvenir production is complex, and is not only dependent on the nature of the tourism industry, but also the social/cultural conditions in the local community. The chapter now turns to an overview of the study area and research methods used in the study.

Study Area and Research Methods

The study area is located in the Kunisaki Peninsula in the Prefecture of Oita on the southern island of Kyushu. Oita prefecture is located approximately 789 km (490 miles) southwest of Tokyo. Even though it is only 1 hour 45 minute flight and 5 hours 15 minutes by the *Shinkansen* bullet train to the nearest major railway station, metropolitan Tokyo and rural Oita are idiosyncratically different in terms of culture, life-style and local people's beliefs and mentality towards life. The rural area tends to hold on to tradition, which includes a strong sense of male/female roles.

Kunisaki Peninsula was created by volcanic eruption, which shaped mountain peaks in the centre of the peninsula and it is surrounded by shallow coast lines (Oita-Press, 2008). Interestingly the soil lacks volcanic-ash-based *andosols*, which turns out to be best suited soil condition for irrigation rice farming. Irrigation farming in the Kunisaki Peninsula has been known since the second century (Nogyo-Kankyo Gijutsu Kenkyusho 2008).

Historically, owing to the geographical and physical isolation of the mountainous peninsula, the Kunisaki communities under investigation tend to display superficial pleasantries to outsiders even though there is a deep-rooted mistrust/distrust of them, which has been documented as an often typical characteristic of a rural community (e.g. Ferrell & Hamm, 1998; Zuehlke, 2008). In order to penetrate the barrier of mistrust, the authors stayed in a community for six weeks to three months at a time, participating in community events and meetings, helping out with neighbourhood chores, and they made formal and informal visits to communities and research sites (e.g. farmer's markets) since 2003. Initially with participatory observation, conversational data collection and more formal interviews, the authors gained a basic understanding of the social and political constructs of the communities. After a few years of regular visits, the farmers and fishermen began to talk to the authors in a more unsolicited manner and they appreciated that the outsiders (authors) were interested in listening to their stories. In most cases the farmers and the fishermen determined the course of conversation and the types of information they provided, however, at times the authors prompted with questions to probe issues further. This informal style seemed more appropriate as the more formal rigid interviews with farmers were less successful.

Even though the authors spoke to any farmers and fishermen in the communities they encountered, there were several people who assumed a position of leader in the community and voluntarily came more frequently to speak to the authors. These key informants (eight people) were usually: in charge of farmers' markets, Bed and Breakfast operators, a leader of women's groups in the communities or instructors of the Organic Farming group. They were in the position to oversee and organise other farmers and fishermen in their groups and therefore they could convey views and opinions about the group members. Local government officials and staff at the local Japan Agriculture organisation who were in charge of the Green Tourism promotion in the region were also helpful in providing detailed information (six people).

The authors also organised three focus groups in 2003–2004. A list of Green Tourism key contacts was given to the authors by the municipal office in Matama. Focus group meetings were arranged with those who were willing to be involved in the research project. In the first focus group, Green Tourism participants in Matama Village were invited to a local community hall to participate in the session. All female farmers in the village who were available at the time of the meeting attended, as well as two male community members who had links to Green Tourism (12 people). The second focus group was with the Green Tourism Steering Committee in Tashibunosho Village. All three male committee members attended the meeting. The third focus group was with the Green Tourism Food Production Women's Group in Kakaji Village. Six female farmers who were working on souvenir food production on the day of the focus group participated in the meeting. What was interesting about this group was that they were all over the age of 65, with the oldest being over 90. The participants of these focus groups maintained communication with the authors in subsequent years. In addition to speaking to the participants, the authors also took photographic inventories of the souvenirs for sale at a variety of farmers' markets and tourist-related shops. The following sections of the chapter will explore the types of souvenirs sold, followed by an examination of issues surrounding female empowerment.

Green Tourism Souvenir Products

Table 8.1 contains examples of souvenirs for sale at a selected number of tourist-related establishments or souvenirs produced by selected groups. A photographic inventory was taken at each location. Two farmers' markets have been included in the table as they are of different scale. The first is Magokoro Farmer's Market, which is a small fruit, vegetable and flower market that also sells souvenirs. It is run primarily by women farmers in the community and sells produce of the size that may not be accepted by Japan Agriculture for the national food terminal distribution system. The second farmers' market is Sun Western, which is larger in scale offering a small restaurant as well as fresh seafood for sale from tanks in the store. Sun Western operates as a cooperative and members pay a fee to display their souvenir products on the shelves. Tour buses often stop here, including an old-fashion heritage bus tour.

SpaLand is a large-scale spa hotel and resort complex. In Japanese culture, visiting the local spa is an important part of the culture and this facility not only attracts domestic and international tourists, but it also attracts a large number of locals who use the spa several times a week. Inside the main lobby

Table 8.1 Examples of souvenirs available

Locations	Food-products	Non-food products and crafts
Magokoro Farmers Market (small scale)	Fresh seasonal food Honey Home-made jams Green tea Dried shiitake mushroom Dried herbs, vegetables, and fruits Organic rice Pickles Salt toffee	Bamboo buckets Elaborated origami crafts Handmade basket with vines Owls (wood or kimono textile) on the branch Crochet dishcloths Buckwheat husk pillow (traditional triangle prism) Recycled Kimono materials • Arm covers • Tissue box case
Sun Western Local Produce Shop (Farmers and Fishermen Market) (large scale)	Honey Home-made jams Green tea Dried shiitake mushroom Dried herbs, vegetables, and fruits Organic rice Pickles Salt toffee	Crochet fish ornament Woven bamboo baskets Recycled Kimono materials • Handbags, tote bags, and purses • Water bottle cooler Flax crochet tote bag Ceramics and pottery
SpaL and Shop	Fresh seasonal food Seafood products (soup mix, dried fish and squid, dried fish roll) Soy bean powder [kinako] products	Bamboo ogre mask Gourd water bottle Gourd ornaments Toys (tops, darts, games) Buckwheat husk pillow

Group	Food products	Craft products
	Honey Home-made jams Green tea Dried shiitake mushroom Dried herbs, vegetables, and fruits Pickles Salt toffee	(traditional and contemporary shapes) Straw sandals Recycled Kimono materials • Square basket • Handbags, tote bags, and purses • Hats Bamboo baskets Wooden cooking utensils, spoons and chopsticks Woven bamboo place mats and coasters Special soaps (mineral soap, sulphur-mud soap)
SpaL and Seniors Workshop	Honey Home-made jams Green tea Dried shiitake mushroom Dried herbs, vegetables, and fruits Organic rice Pickles Salt toffee Rice	Ceramic and pottery Recycled material sandals Recycled material string bags, hats Buckwheat husk pillow (contemporary shape) Traditional quilted embroidery Elaborated origami ornaments
B&B in Tashibunosho		Rice husk turtle and crane
Tashibunosho Women's Group	Bento-box made from seasonal vegetables and seafood	
Kakaji Women's Group	Fried rice crackers Salt candy Vegetable preserves	

of the resort is the souvenir shop selling local food products and souvenirs. On the resort property the owners have set up various other small stand-alone shops and one of these is the SpaLand Seniors' Workshop. This workshop is staffed by seniors and they also make the souvenir products.

A Bed and Breakfast is included in Table 8.1 as a location where the owners made traditional souvenirs (ornamental turtles and cranes) out of rice husks for their guests. Finally two Women's Groups are included in Table 8.1 from two different villages and they focus on producing food-related souvenirs.

Table 8.1 is divided between food products and non-food products and crafts. As the Kunisaki Peninsula is surrounded by an inland sea, foodstuff from the sea as well as agricultural products are the main food products. Most of the food products are in the form of preserved food, mostly dried food products and pickles. However, with changes in customer demographics and preferences, the farmers are adopting new food products such as jam, preserves and baked goods.

At SpaLand, fresh bread is made by a local group and delivered daily for sale in the shop. A picture of the group is displayed above their products. One woman also started her own bakery and delivers bread and bake goods to SpaLand. Fresh seasonal produce is also available for sale. The Tashibunosho Women's Group worked together to produce bento-boxed (Japanese style lunch boxes) lunches that were made from seasonal vegetables and seafood. These were delivered and sold to tourists at the local museum and a shop in the Heritage Town district of Bungo-takada city.

The authors had the opportunity to attend a working meeting of the Kakaji Women's Group who were in the process of making fried rice crackers, which were packaged and sold to tourists through local farmers' markets. Innovation is also shown in packaging and product development. The packaging of the food products ranges from quite simple clear plastic bags with the producer's name on it, to very high quality decorative wrapping with designs and pictures. High-end packaging gives the impression of high quality and price value for simple local products. For example, tea and duck rice mix are packaged in airtight packages. The Kunisaki Peninsula duck is famous and duck rice is a simple dish yet known as a local delicacy. As the product is perishable, cooked duck rice cannot be sold in the shops, therefore they produced duck rice mix which people can add to rice and cook.

Both men and women make the traditional arts and crafts sold as souvenirs during the winter or the seasons between the more arduous fieldwork seasons. Many of the items are of daily necessity (such as bamboo water buckets, recycled material tote bags, straw sandals, etc.) rather than ornamental souvenirs. Owing to the location of the villages studied, the types of souvenirs produced are rather traditional, old fashioned and unsophisticated. Considering the fact that the producers are aged people, who use their spare time, particularly in the wintertime to create the souvenir goods, what they produce is not so different from what they produce for their own use. Traditional arts and crafts in this region come from bamboo materials, vines and agricultural waste products, such as straw and husks. Thrifty farmers tend not to waste re-usable textiles, such as old kimonos and clothing, and these are re-used in souvenirs. Recycled Kimonos have been turned into handbags, tote bags and purses, while bamboo was used to make cooking utensils.

Basket weaving, bamboo crafts and vine crafts are all disappearing art forms in Japan and very few people know how to do this intricate work. As a more modern convenient life style has evolved, nature-based elements, such as straw and buckwheat husks, have been discarded in quantity. However, with the recent trends of rising awareness of environmental conservation and preservation of traditional arts in Japan, buckwheat husk pillows and straw sandals have seen new uses for these formerly discarded natural elements. The positive health effects of buckwheat husks in pillows have been featured in various health magazines.

At several of the shops straw sandals are sold and they are based on the revival of traditional rush (straw) farming. A husband and wife team started this project in 1994 and a group of farmers recently joined in producing high-quality rush for *tatami* mats. Demand for *tatami* mats, which are used as flooring material in traditional Japanese houses, has been in decline from the 1980s, however, this group's product is still in high demand nationwide and they developed a line of straw sandals in 2001. The product has moved beyond small-scale souvenir production, as they now have orders from national department stores and from Germany for the sandals. Even though this is a successful case of entrepreneurship, their profits remain modest.

Ceramic pottery is also produced for the SpaLand Seniors' Workshop. The ceramics included more traditional tea sets and bowls, as well as more modern-looking vases and plates.

The types of souvenirs sold can also be analysed by Gordon's (1986) typology of souvenirs, but it was found that the Green Tourism souvenirs only fit into the category of 'Local Products'. The local Green Tourism products that are food-based have become well established as part of regional cuisine and it is now strongly linked to the identity of the area. These products take on some of the attributes of becoming a 'Marker' in Gordon's typology owing to the strong association of place. Hashimoto and Telfer (2007) examined souvenirs in the Niagara region for evidence of geographically displaced authenticity and when applying this concept to rural Kunisaki, there was no evidence of geographically displaced authenticity as the products are all emphasising the local environment/destination.

Souvenirs and Female Empowerment

Of the various markets, shops and groups investigated, many of those involved in the production and sale of souvenir items were women. Life expectancy and an aging population in these communities left many more elderly women than men to take part in Green Tourism projects. Participating in these projects provides more social networking opportunities and socialisation events to otherwise isolated elderly people. The small income from the sale of souvenirs is nevertheless important to these elderly people who do not receive sufficient pensions, yet the socialisation aspect of getting involved in the arts and crafts making seems to be very rewarding.

The authors attended a Kakaji Women's Group working session and it was clear that there was an important social component for these women. Making the souvenirs together in a group brought the women together where they could meet old friends while working on a common project. It was interesting to note that many of the women interviewed were over 70 years of age. The elderly people also believe using their hands and fingers in making souvenir products helps keep their mind young and active, which is supported neurological studies (e.g. Chiarenza *et al.*, 1991; Restak, 2009). The SpaLand Seniors' Workshop was also the centre of community gatherings for seniors, and many of the people working at the shop selling the souvenirs were women. It was a place where the souvenir producers could stop by and talk, and was an important part of their social network and promotion of wellbeing. What these women have been producing in the past and may have been taken for granted by family members, now has added value.

The regional government has been urging villages to come up with innovative arts and crafts using waste materials, such as the buckwheat husks mentioned above, which can be sold as souvenirs. The city of Bungo-takada hosted a seminar to teach women how to make the buckwheat husks pillows and over 60 women attended. The workshop was a result of a request of the Bungo-takada buckwheat producers' union to find a way to recycle the buckwheat husks. The pillows are sold at farmer's markets, SpaLand and also in a heritage tourist district. The workshop not only has the obvious technical training component, but it is also has a social dimension for the participants.

As the research area has focused on Green tourism in rural areas, there are several challenges to the souvenir industry that relate to farming and many of these impact the woman producers. One of these issues is the time conflict with traditional farming. The woman farmers that supplied souvenir products to the various farmers' markets are faced with not only their work on the farm, but also with the domestic work that is often attributed to women. A typical day for a woman farmer in the rural area under investigation involves rising early, making their husband his breakfast, followed by joining her husband in the field. The woman farmer will make deliveries to the markets and at the end of the day will be required to make dinner for her husband, as well as being responsible for the domestic chores including cooking, cleaning and laundry. This often leaves the women farmers facing difficulty in finding enough time to make the souvenirs (see also Hashimoto & Telfer, 2011).

The seasonality of tourism and farming can also coincide, so when life is busy on the farm in the summer the number of tourists can also rise, which places additional responsibilities on the women farmers involved in the tourism industry. This is especially the case for women farmers who also run Bed and Breakfasts. In order for the women to be successful in the production of souvenirs there needs to be support from family members to take time away from farming or domestic chores to make and deliver the souvenir products. This finding is similar to the point being argued by Swain (1993) noted above, that the internal (family/community) factors were very important with regards to woman empowerment through tourism.

Conclusion

In many rural areas going into decline, tourism is often turned to as a means for economic rejuvenation. Through the development of Green Tourism products in the Kunisaki Peninsula encouraged by the Japanese Ministry of Agriculture, Forestry and Fisheries, there has been an increased role for women to play in the development of souvenirs for tourists. The government support focuses more on the technical support, such as workshops, however financial support is limited to initial/start-up finding. Opportunities exist to expand the line of souvenirs as well as the locations they are displayed and sold. Yet, historical mistrust of outsiders does not help to forge effective networks, but tends to reinforce the territorial barriers and limits the potential expansion of markets. On the other hand, this territorial barrier of not collaborating with other villages seems to maintain the regional authenticity of souvenir goods. Regional authenticity is also maintained as Green Tourism is not the main source of income and so the sales of souvenirs at the cost of displaced authenticity (introducing souvenir products from other regions) is not likely to occur.

One of the main challenges for the remote rural areas in Japan is the aging of the population and so the scope of the projects in these areas will be limited. The souvenir goods they produce are simple, old fashioned and traditional, and these may not necessarily be attractive to all tourists. While the Green Tourism souvenirs do not bring in a great deal of money owing to the scale of the projects, the women can earn sufficient pocket money for their own use. One of the key findings is that the souvenir projects do help generate a sense of self-worth and self-value, in part through the recognition of what they have been making has value to outsiders. Being traditionally employed in isolated farm work, the production of souvenirs has an important socialisation element and allows the women to get together with friends off the farm.

References

Apostolopoulos, Y., Sönmez, S. and Timothy, D.J. (2001) *Women as Producers and Consumers of Tourism in Developing Regions.* Westport, Prager.

Blundell, V. (1993) Aboriginal empowerment and souvenir trade in Canada. *Annals of Tourism Research* 20, 64–87.

Chiarenza, G., Hari, R., Kahu, J. and Tessore, S. (1991) Brain activity associated with skilled finger movements: Miltichannel magnetic recordings. *Brain Topography* 3 (4), 433–439.

Cohen, J. (2001) Textile, tourism and community development. *Annals of Tourism Research* 28 (2001), 378–398.

Cone, C. (1995) Crafting selves: The lives of two Mayan women. *Annals of Tourism Research* 22 (2), 314–327.

Connelly-Kirch, D. (1982) Economic and social correlates of handicraft selling in Tonga. *Annals of Tourism Research* 9, 383–402.

Costa, J.A. (2005) Empowerment and exploitation: Gendered production and consumption in rural Greece. *Consumption, Markets and Culture* 8 (3), 313–323.

Davis, C. (2007) Can developing women produce primitive art? *Tourism Studies* 7 (2), 193–223.

Ferguson, L. (2011) Promoting gender equality and empowering women? Tourism and the third Millennium Development Goal. *Current Issues in Tourism* 14 (3), 235–249.

Ferrell J. and Hamm, M.S. (1998) *Ethnography at the Edge: Crime, Deviance, and Field Research.* Boston: Northeastern University Press.

Gordon, B. (1986) The souvenir: Messenger of the extraordinary. *Journal of Popular Culture* 20, 135–146.

Hashimoto, A. (2002) Tourism and social-cultural development issues. In R. Sharpley and D.J. Telfer (eds) *Tourism and Development: Concepts and Issues* (pp. 202–230). Clevedon: Channel View Publications.

Hashimoto, A. and Telfer D.J. (2007) Geographical representations within souvenirs in Niagara: The case of geographically displaced authenticity. *Tourism Geographies* 9 (2), 191–217.

Hashimoto, A. and Telfer, D.J. (2011) Female empowerment through agritoursim in rural Japan. In R. Torres and J. Momsen (eds) *Tourism and Agriculture, New Geographies of*

Consumption, Production and Rural Restructuring (pp. 72–84). Abingdon, Oxon: Routledge.

Hitchcock, M. and Kerlogue, F. (2010) Tourism, development and batik in Jambi. *Indonesia and the Malay World* 28 (82), 221–242.

Markwick, M. (2001) Tourism and the development of handicraft production in the Maltese islands. *Tourism Geographies* 3 (1), 29–51.

Nason, J. (1984) Tourism, handicrafts, and ethnic identity in Micronesia. *Annals of Tourism Research* 11(3), 421–449

Nogyo-Kankyo Gijutsu Kenkyusho [National Institute for Agro-Environmental Sciences] (2008) *Kazankoku Nippon to Dojou Hiryou-gaku* [Volcano Nation Japan and Soil-fertilisation Study]. Nogyo to Kankyo No.103 (1 November 2008). Available from http://www.niaes.affrc.go.jp/magazine/103/mgzn10306.html

Oita-Press (2008) *Oita Isan* [Oita Heritage] *Kunisaki hantou no yamano rekishi to keikan* [history of the mountains and landscape of Oita peninsula] (1 April 2008). Available from http://www.oita-press.co.jp/featureNews/120894237407/2008_1208944332.html

Restak, R. (2009) *Think Smart: A Neuroscientist's Prescription for Improving Your Brain's Performance*. New York: Riverhead Books.

Swain, M. (1993) Women producers of ethnic arts. *Annals of Tourism Research* 20 (1), 32–51.

Swanson, K.K. and Timothy, D.J. (2012) Souvenirs: Icons of meaning, commercialisation and commoditization. *Tourism Management* 33 (3), 489–499.

Zuehlke, E. (2008) Empowering women's voices on reproductive health in the media; and taking stock of reproductive health and the media. Available from http://www.prb.org/Articles/2008/womenseditiontraining.aspx?p=1

9 Understanding Tourist Shopping Village Experiences on the Margins

Laurie Murphy, Gianna Moscardo and Pierre Benckendorff

Tourism is often encouraged by governments and development agencies as a way to support traditional economic and social activities in regional areas, especially peripheral or marginal places. Many small villages on the margins have turned to tourism as a development option and have pursued, either deliberately or by serendipity, a strategy of offering tourist shopping as a way to support local production of arts, crafts and specialist food and beverage (Murphy *et al.*, 2011a). Small villages on the margins that provide retail outlets have been referred to as Tourist Shopping Villages (TSVs) (Getz, 2000). But there is little consistent evidence that the development of tourist shopping brings benefits for local residents and/or producers. This chapter will examine the challenges that tourism presents to these TSVs, focusing on the links between the availability of locally produced souvenirs, positive tourist experiences and the maintenance of support for local souvenir production and sales. In particular the chapter will analyse the nature of shopping experiences in these villages on the margin, the links between these experiences and consumption of locally produced souvenirs.

Wilkins (2011) provides an overview of research into souvenirs, and in that work offers several definitions, concluding that souvenirs are purchases made by tourists that can act as tangible evidence of tourist experiences, as aids to the recollection of experiences, as gifts for self and others and as symbols that give meaning to tourist experiences. From this description it is clear that any number of things can be seen as souvenirs, but typically they are objects with a close connection to the place visited. For the purposes of this chapter, souvenirs will be defined as any product purchased by tourists for other than utilitarian purposes, which has a physical or symbolic connection to the destination. These can include food, beverages, arts, crafts, homewares, apparel, toys and books (Collins-Kreiner & Zins, 2011; Wilkins, 2011). The focus in this chapter is, therefore, particularly on tourists purchasing locally produced souvenirs.

Most of the research into souvenirs has analysed the ways in which souvenir production evolves over time and the interaction between destination culture, tourism development and souvenir design focusing on issues of authenticity and commodification (Swanson & Timothy, 2012). In this chapter we are more concerned with how the production, presentation and sale of souvenirs can assist local communities in peripheral regions. The authors have been involved in a series of research projects conducted in more than 50 TSVs in Australia, Canada, New Zealand, the United States, England and Ireland (see Murphy *et al.*, 2008, 2011a, 2011b) and these studies provide evidence that:

- it can be difficult to get tourists to actually buy products while visiting the villages;
- pressures from tourism growth can also make it difficult to retain locally produced arts, crafts, and food and beverages as the core products sold in the village; and
- key dimensions of the tourist experience are likely to have significant impacts on visitors' souvenir purchasing.

To better understand the factors that support tourist consumption of locally produced souvenirs in TSVs, this chapter will provide a review of the relevant literature focussing on the evolution of TSVs and implications for both the nature of the tourist experience and consumption of local souvenirs. It will then examine in more detail evidence from surveys conducted at two Australian TSVs, Hahndorf and Montville. An analysis of the connection between tourist experience and purchase, conducted specifically for this chapter, identified a number of features of TSV presentation and management that contribute to a greater likelihood of a souvenir purchase, and these results will be presented. The chapter concludes by identifying key dimensions of the village experience that can encourage tourists to purchase local souvenirs and support local communities on these margins.

Literature Review

Two key traditions of research are relevant to the present discussion – studies into the evolution or development of TSVs and studies linking features of tourist shopping experiences to the purchase of souvenirs. Studies into TSV development pathways reveal some consistent patterns and Murphy and colleagues (2011a) provide a detailed description of two of these patterns. In their study, conducted with key stakeholders in a number of TSVs in the US New England region, the interviewees identified a number of changes in the TSVs resulting from tourism growth and amenity migration. In the first development pattern these two processes were related, with amenity migration partly driven by the availability of business and employment generated by tourism development, and in turn fuelling development and change in tourist retailing opportunities. Growth in tourist numbers and/or resident population increased the range of ancillary services, like accommodation and restaurants, and the value of existing businesses. Landlords, encouraged by increased revenue and business value, often raised rents and local business owners sold to other external parties taking advantage of the rising real estate values. The rise in property prices then made it difficult for local producers and artisans to remain in the village and they were often replaced by larger operations selling more generic souvenirs. Alternatively, in the second development pattern the increased tourist and resident numbers, combined with low land prices, encouraged outlet mall developers and a rise in 'big box' and budget retail options.

The people interviewed in this study were divided in their responses to these evolutionary paths. For some, tourism development offered economic and social opportunities that enhanced their quality of life, and they believed that the outlet malls and more generic and budget shopping options attracted visitors who also supported smaller local businesses. Others felt that the changes in retail options attracted a different type of tourist who was not interested in regionally distinctive souvenirs. In all of these discussions there was an assumption that changes in retail options altered the nature of the experience available to the tourists and consequently affected support for locally produced goods and services. These development cycles have also been described and linked to the concept of creative destruction in a number of other TSVs in Canada and China by Mitchell (2003) and colleagues (Mitchell & de Waal, 2009; Fan et al, 2008).

What is important to note in these discussions of TSV evolution is the lack of evidence about the nature of the tourist shopping experiences and how this is linked to actual purchases. Many assumptions are made about how the changing nature of the shops and the village atmosphere is likely to influence tourist behaviour, but very little evidence exists to test these assumptions. There is, however, some evidence available from research into leisure shopping in general. Reviews conducted by Turley and Milliman

(2000), Machleit *et al.* (2005) and Underhill (2004), for example, identify three sets of factors that can make a significant contribution to positive shopping experiences.

- Ambient variables, such as air quality, cleanliness, comfort and lighting.
- Design variables, such as signage, window displays, payment options and variety and quality of merchandise.
- Social variables, such as crowding, staff attitudes and social interactions.

While these shopping experience factors have been consistently linked to overall experience evaluations there is much less evidence linking them to purchase behaviour because very few studies have measured actual buying behaviours. Most research focuses on intention to purchase, which has been shown to be positively related to actual purchase behaviour (Tsiotsou, 2006). Research conducted into actual purchase behaviour suggests that there is a positive link between satisfaction with shopping experience factors and actual buying behaviour (LeHew & Wesley, 2007; Zhuang *et al.*, 2006). This research also shows that the levels of actual shopping expenditure for leisure shoppers is, however, often lower than expected by store owners and managers because the enjoyment for leisure shoppers lies mainly in the process of shopping not solely in the purchase of the goods (Jones *et al.*, 2006; LeHew & Wesley, 2007; Zhuang *et al.*, 2006). This is consistent with findings reported by Murphy and colleagues (2008), where business owners and government officials participating in workshops conducted in four Australian TSVs stated that many tourists browse, but very few actually shop.

In summary, there is evidence that changes designed to encourage the touristic appeal of TSVs can have a number of unintended negative impacts on the residents of these villages. These changes can discourage both the production and presentation for sale of local souvenirs, as well as tourist purchases of local souvenirs. It seems that in some places the development of tourist attractions, festival and events programmes and street entertainment can support a positive tourist experience but leave little attention left for shopping of any sort. Understanding the links between dimensions of the shopping experience, with a particular focus on features linked to local souvenirs, overall satisfaction with the TSV experience, and purchase behaviours and intentions, is important in supporting the success of strategies to use tourist shopping for local souvenirs as a way to support traditional activities and economies of places on the margins. The study presented in the following sections attempts to address this gap in our understanding of tourism shopping.

The research setting

Both Hahndorf and Montville are small rural villages on the periphery of the urban centres of Adelaide and Brisbane, respectively. Hahndorf was settled by German settlers in the late 1840s and German heritage is expressed through food, wine and art and forms a major part of the tourism promotion for the town (Murphy *et al.*, 2011a). In addition, the town has 22 state heritage listed buildings and has since the 1970s become a major tourist destination (Department of Heritage and Environment, 2009). Montville is located northwest of Brisbane, in the hinterland of a major coastal tourism destination, the Sunshine Coast. Montville grew as a service centre for forestry and dairy farms, but, like many other rural areas in Australia, experienced major changes in the 1970s and 1980s as a result of a decline in traditional agricultural activities. A number of alternative lifestyle groups moved into the surrounding area attracted by the possibility of buying cheap but productive agricultural land to support communes, the access to natural environments and the warmer climate (Carter *et al.*, 2007; Metcalf, 1995). The production of arts and crafts by these residents created new economic opportunities and led to the rejuvenation of Montville as a mountain rainforest-themed centre with subsequent tourist shopping development (Murphy *et al.*, 2011a).

These two TSVs share many of the characteristics of similar villages studied in the United States, Canada, the United Kingdom and New Zealand, including being located in rural areas on the periphery of major urban or tourist centres, a focus on heritage buildings and a heritage streetscape as a key attraction, extensive street and landscaping, and a variety of souvenir, arts and craft stores supplemented by cafes and restaurants. Hahndorf has a clear and consistent German heritage theme and a number of longstanding local producers of food and wine, and so has been able to retain many local products in its shops. In recent years, however, there has been a decline in the provision of local products and an increase in shops specialising in more luxury-branded goods. Montville has had a less coherent theme with many new buildings in an eclectic mix of architectural styles and shops providing a range of goods but little specifically produced in the local region. The use of these two villages thus provides differing village experiences and the opportunity to explore links between tourist consumption and experience across a range of dimensions.

Study Methodology

In order to better understand the nature of the tourist experience in TSVs, more than 700 tourists were surveyed in the two villages with approximately

500 surveys collected in Hahndorf and 200 in Montville. The surveys measured details of the types of purchases made, the amount spent, and the likelihood of purchasing souvenirs once back home, as well as evaluations of the TSV experience and shopping opportunities, motivations, expectations and personal characteristics. The TSV experience dimensions were measured in two parts – elements or features of the shops within the village and elements of the overall TSV experience. In each case tourists were asked to rate both the importance of each dimension to them and the performance of the village and the shops they visited.

The surveys were collected at main transit nodes within the villages and were conducted during two weeks in the peak tourist season and included weekdays, weekends and school holidays (see Murphy *et al.* (2011b) for further method details). The resulting sample is profiled in Table 9.1 and provided data from a range of different tourists.

Results and Discussion

Patterns of souvenir purchase

The surveys conducted in these two TSVs included questions that asked respondents to estimate their expenditure in a number of different categories. Table 9.2 summarises the patterns of responses to these questions and shows that a considerable proportion of visitors do not spend very much in the villages beyond food and beverage, with more than half not purchasing anything from the local shops and the majority of those that do purchase from these local shops spending less than AUD$30 or less. This is consistent with LeHew and Wesley's (2007) research on tourist behaviour in shopping malls or centres.

The sample of tourists were asked to rate their likelihood of: buying products from the TSV if they were available for purchase in stores at home (28% very likely for Hahndorf, 8% for Montville); ordering products directly from TSV suppliers (10% very likely for Hahndorf, 5% for Montville); and recommending products from the TSV to friends and family (47% very likely for Hahndorf, 23% for Montville).

Three important points need to be made about these figures on post-visit purchase behaviour. First, these are measures of intended rather than actual behaviour. Second, the differences between the two TSVs reflect differences in opportunities in that there are a number of foods and wines from Hahndorf available in stores nationwide making it easier to engage in post-visit consumption. Third, there are opportunities through souvenir

Table 9.1 TSV visitor sample profile

Variables	Sample profile
Male	39%
Female	61%
Mean age	45 years
<30 years	28%
31–50	36%
>50 years	36%
Usual place of residence	
• Australia	86%
• Overseas	14%
Visited the village previously	72%
Length of stay away from home	
• Day trip	45%
• 2–3 days	11%
• 4–7 days	15%
• 7 days	29%
Travel party	
• Alone	6%
• Spouse/partner	41%
• Family	29%
• Friends	19%
• Tour or other group 5%	5%

production and retail to use tourists to support the activities and econo-mies of these peripheral villages. For example, when those who said they were likely to engage in post-visit purchase or recommend purchases were asked what they would be likely to purchase, the six most common options were food (62%), craft (41%), wine (37%), art (28%), homewares (12%) and

Table 9.2 Summary of expenditure

Type of expenditure (n = 702)	% who report expenditure in category	Mean expenditure of those who did
Food and drink	64%	AUD$43
Purchases from shops	46%	AUD$30
Accommodation in/near TSV	13%	AUD$250

Table 9.3 Correlations between overall satisfaction, TSV performance and expenditure

	Score on total performance on TSV dimensions	Overall satisfaction	Expenditure in the shops
Score on total performance of the shops within the village	0.54		
Score on total performance on TSV dimensions		0.48	
Overall satisfaction			0.12

Figures are Kendall's Tau-B and all are significant at p < 0.001. The total performance scores are regression factor scores derived from single factor principal components analyses.

clothing (12%). All of these categories are possible options for supporting communities on the margins.

These opportunities offered by TSVs were also supported by the finding that there were significant positive correlations between overall satisfaction with the TSV experience and ratings of how well the TSV performed on various dimensions, and measures of expenditure (see Table 9.3). The correlations presented in Table 9.3 confirm overall links between performance on shopping experience dimensions, a positive tourist experience and an increased likelihood of buying souvenirs in the TSV.

Linking satisfaction, shopping experience and souvenir purchase

To further examine the details of these relationships, the tourists in the sample were organised into four shopper groups based on propensity to purchase or recommend local products post-visit and whether or not purchases were made on site. The resulting four groups were labelled 'non-shoppers' (29%) who did not purchase any souvenirs on site and had low ratings for all the post-visit purchase variables; 'local only shoppers' (27%) who did purchase souvenirs on site but had low ratings for the post-visit purchase variables; 'post-visit only shoppers' (20%) who did not purchase on site but who gave high ratings on the post-visit purchase variables; and 'local and post-visit shoppers' (24%) who were both consumers on site and post-visit.

Once these groups were established they were then compared and profiled on a number of other variables. It could be argued that shopping behaviour is linked to features of the individual tourist rather than the TSV experience. For example, results of other studies on shopping involvement and motivation showed that shopping enthusiasts are more likely to be female (Arnold & Reynolds, 2003; Guiry et al., 2006; Josiam et al., 2005). To explore this possibility the four shopper groups were compared on a series of socio-demographic, travel and general leisure shopping behaviour variables. The statistical tests indicated that there were no significant differences across the four shopper groups with respect to age, place of origin (Australia versus international), travel party, previous visits to the village, regularity of leisure travel shopping or enjoyment of shopping at shopping malls, urban precincts or duty-free stores. There were, however, differences between the four groups for gender ($\chi^2 = 9.531$, $p = 0.023$), with a greater percentage of females in the purchase groups (50% of non-shoppers, 59% of post-visit only, 60% of local only and 69% in local and post-visit) and the importance of shopping in travel decisions – with local only (10.2%) and local and post-visit shoppers (14.3%) less likely than non-purchasers (20.3%) and post purchase only (17.9%) to say they would avoid shopping. Zhuang et al. (2006) highlighted the importance of being motivated to shop in predicting actual consumption behaviour.

The analysis of the importance of shopping in travel decisions also revealed that 44% of those who stated that shopping was the main reason for travel and 48% of those who stated shopping was an important factor in travel decisions did not make any purchases during their visit. These results suggest that while there are tourists with a strong personal interest in and desire to shop, their actual purchase behaviours depend, at least in part, on the opportunities and experiences offered in the villages and that there is potential to encourage greater consumption of locally produced souvenirs and products.

Existing research into the evolution of TSVs highlighted the question of how tourists interested in local and regionally distinctive souvenirs would respond to other retail opportunities, especially direct outlet malls or centres. To explore this issue, the survey included a question asking tourists how much they enjoyed shopping in a range of other settings. There were no significant differences in the responses of the four shopping groups for enjoyment of duty-free shopping, urban shopping precincts or traditional shopping malls or centres. There were, however, differences for shops in and around tourist attractions ($F = 7.172$, $p = 0.000$), markets ($F = 9.460$, $p = 0.000$), direct outlet shopping ($F = 3.947$, $p = 0.008$) and shopping in small villages and towns ($F = 17.553$, $p = 0.000$) – these are summarised in

Table 9.4. Some of each of the three shopper groups enjoyed all types of shopping, including outlet mall, while others were more exclusively focused on local and regionally distinctive opportunities. Clearly the development of outlet and more generic retail in a TSV would change the overall profile of the tourist arrivals, but it is not clear that this would diminish demand for locally produced goods.

The four groups also differed significantly on the length of time spent in the village ($F = 3.209$, $p = 0.023$) with higher mean hours for the two on-site shopping groups (mean of 2.5 hours for non-shoppers, 2.2 hours for post-visit only, 2.7 hours for local only and 3.1 hours for local and post-visit). Yuksel (2007) provided evidence that longer time spent in a shopping precinct was related to greater intention to purchase and then argued that, in turn, this would lead to greater actual purchase. These results support that argument.

More detailed analyses of the responses of the four shopper groups to the importance of features of the village as a whole and the souvenir shops more

Table 9.4 Enjoyment of other shopping opportunities by TSV shopper groups

Other shopping types	Non-shoppers	Post-visit only	Local only	Post-visit and local
In/around tourist attractions *(n = 453)*				
Mean* ($F = 7.17$, $p = 0.000$)	3.1	3.6	3.2	3.7
% Enjoy not at all	11	5	11	6
% Enjoy very much	9	23	14	30
Markets				
Mean ($F = 9.46$, $p = 0.000$)	3.8	4.2	4.0	4.4
% Enjoy not at all	4	1	3	1
% Enjoy very much	30	48	40	58
Direct outlet				
Mean ($F = 3.9$, $p = 0.008$)	2.8	2.9	2.8	3.4
% Enjoy not at all	22	22	20	17
% Enjoy very much	11	17	15	30
Small villages and towns				
Mean ($F = 17.6$, $p = 0.000$)	3.7	4.2	4.0	4.5
% Enjoy not at all	22	22	20	17
% Enjoy very much	11	17	15	30

* Mean based on rating scale from 1 (not at all) to 5 (very much)

specifically, found a number of significant differences between the groups. An investigation of where the differences lay provides insights into links between purchase of local souvenirs and TSV experience. In terms of the overall features of the village, all three types of shoppers were similar and gave significantly greater importance than the non-shoppers to the quiet day out ($F = 3.821$, $p = 0.010$), availability of products made in the local area ($F = 8.756$, $p = 0.000$) and/or products unique to the region ($F = 9.296$, $p = 0.000$), the opportunity to meet locals ($F = 10.386$, $p = 0.000$), overall village atmosphere ($F = 10.539$, $p = 0.000$) and being able to see local products being made ($F = 8.362$, $p = 0.000$). The post-visit only and post-visit and local shoppers shared an interest in escaping the city ($F = 11.786$, $p = 0.000$), attending a special event ($F = 7.765$, $p = 0.000$) and learning about the place ($F = 6.456$, $p = 0.000$).

In addition to rating the importance of village features, the survey respondents were asked to select the six most important things that a successful TSV requires. Table 9.5 provides a summary of the responses from the four shopper groups. While not statistically significant, the pattern of responses revealed that higher levels of consumption were related to a desire for variety, local heritage and ease of access and orientation.

Finally, there were significant differences between the four shopper groups in the importance ratings given to features of the shops within the village. In particular, the three shopper groups differed from the non-shoppers in seeking: variety of product selection ($F = 21.897$, $p = 0.000$), sales staff with a positive attitude (6.815, $p = 0.000$), value for money ($F = 17.058$, $p = 0.000$) and regionally distinctive souvenirs ($F = 13.814$, $p = 0.000$). The post-visit and local shoppers were distinctive also in seeing cleanliness ($F = 11.139$, $p = 0.000$), window displays ($F = 9.196$, $p = 0.000$), staff language ability ($F = 6.147$, $p = 0.000$), sales staff efficiency ($F = 7.837$, $p = 0.000$) and payment options ($F = 10.676$, $p = 0.000$) as being very important.

Table 9.5 TSV features more highly desired by different shopper groups

Non-shoppers	Post-visit only	Local only	Post-visit and local
Easy to park	Pleasant	Variety of eating places	Easy to get to
Pedestrian friendly	Landscaping	Visually appealing	Easy to find your
Tourist attractions	Convenient	architecture	way around
Not too crowded	restrooms		Variety of shops
Festivals	Places to rest		Heritage buildings
			Markets

Contributions

This study adds to the existing literature on shopping in TSVs in two main ways. First, it examines actual purchase behaviour and not just purchase intentions. Very few studies in either leisure shopping in general, or tourist shopping in particular, have measured and analysed actual purchase behaviour in any detail. Second, very few studies have examined the tourist experience in TSVs, with most published research on TSVs focused on impacts and consequences for residents, and most research on tourist shopping conducted in settings other than TSVs. This combination of a detailed analysis of tourist shopping experiences and actual buying behaviour provides a unique opportunity to explore the potential for TSVs to generate support for local production and traditional activities. Of particular interest to the theme of this book is the link between tourist shopping experience and purchase behaviour, because it is this link that needs to be established if souvenirs are to play a role in sustaining and preserving craft traditions in communities on the margins.

In this study, five key features of the tourist experience were found to be associated with a higher likelihood of purchasing while visiting TSVs, as well as higher levels of interest in purchasing goods after the visit.

(1) *Village atmosphere:* shoppers preferred a pleasant setting and positive heritage atmosphere. These features of the TSV as a whole also applied to the shops within the village, with shoppers giving higher importance ratings to cleanliness and attractive window displays. This overall atmosphere could be a seen as the most basic prerequisite for TSV shopping.

(2) *Convenience:* shopping was more likely when the respondents thought the TSV was easy to get to and navigate around and had shops that offered efficient sales staff and a range of payment options.

(3) *Variety:* shoppers expressed an interest in a variety of shops, places to eat and merchandise within the shops. This emphasis on variety could be an important one in supporting a range of different local producers and encourages the retention of smaller, specialist stores over fewer but larger generic outlets.

(4) *Experiential activities:* shoppers were significantly more interested in opportunities to see local products being made, to meet local people and to learn about the TSV. This could be referred as 'shopping as local interpretation'. The idea that shopping is a way for tourist to engage with local residents and to learn about the culture and environment of

the destination has been suggested on other studies of tourist behaviour in small villages and at local markets (Carmichael & Smith, 2004; Moscardo, 2004).

(5) *Purchase opportunity:* tourists can only buy locally produced goods and merchandise if these are available in the TSV shops. The three shopper groups all gave high importance ratings for the availability of regionally distinctive and/or locally produced goods. Opportunity was also an important aspect of post-visit purchase. There are a number of barriers that could discourage tourists from on-site purchasing, such as a limited ability to carry purchases while still travelling, concerns over suitability and sizes of souvenirs bought as gifts, time constraints or a desire to focus on other aspects of the experience while on site, such as family bonding, rather than shopping. The opportunity to purchase goods from the TSV region after the visit is both a way to assist tourists to overcome these barriers and an additional benefit from tourism in the promotion of locally produced goods. Options such as online and mail ordering, and information on places where local brands are available outside the TSV, can support these options. Potential exists to increase tourist consumption post-visit and this may be an important mechanism for supporting the viability of local producers and responding to fluctuations in tourist numbers.

Conclusions

In conclusion, TSVs do attract visitors with an interest in the sort of consumption that can support local businesses, craftspeople and suppliers, but there is considerable potential to increase this consumption to provide more income for local communities, and especially those on the margins. Second, there is a strong positive link between an enjoyable tourist experience in a TSV and the likelihood and extent of tourists purchasing local souvenirs.

The study reported in this chapter also has implications in the broader context of the way in which TSVs have evolved and changed over time. It provides some evidence that existing TSV visitors who seek and purchase local souvenirs, may find additional shopping opportunities attractive, and that a move to things like outlet shopping may attract additional tourist numbers to visit towns on the periphery, with an increase in support for local goods. But such developments need to be seen and managed as additions rather than replacements, with careful planning required to ensure that the TSV retains its local heritage, with spatial separation likely to be the best option. This conclusion highlights the need to find new approaches to tourism and town planning that support the retention of local and regionally distinctive souvenirs as the core of the TSV experience, without restricting the ability of local producers to benefit from increased revenues and business value.

It is clear that tourists, on the whole, expect and seek locally made and regionally distinctive souvenirs. This means that strategies for retaining local artisans and producers need to be considered. In addition TSV communities and stakeholders should also explore the potential of post-visit consumption options. The development of online retail provides new opportunities for post-visit consumption, but there are also examples of local products being commercialised for sale in delicatessens, gift shops and supermarkets, department stores and airport concessions beyond the peripheral region. Lastly, retailers and village entrepreneurs should not lose sight of the importance of experiential activities and the idea of shopping as interpretation of place.

References

Arnold, M. and Reynolds, K. (2003) Hedonic shopping motivations. *Journal of Retailing and Consumer Services* 79, 77–95.

Carmichael, B. and Smith, W. (2004) Canadian domestic travel behaviour: A market segmentation study of rural shoppers. *Journal of Vacation Marketing* 10 (4), 333–347.

Carter, J., Dyer, P. and Sharma, B. (2007) Dis-placed voices: Sense of place and place-identity on the Sunshine Coast. *Social & Cultural Geography* 8 (5), 755–773.

Collins-Kreiner, N. and Zins, Y. (2011) Tourists and souvenirs: Changes through time, space and meaning. *Journal of Heritage Tourism* 6 (1), 17–27.

Department of Environment and Heritage (2009) Hahndorf state heritage area. Accessed May 2010 http://ww.environment.sa.gov.au/heritage/shas/sha_hahndorf.html

Fan, C., Wall, G. and Mitchell, C.J.A. (2008) Creative destruction and the water town of Luzhi, China. *Tourism Management* 29 (4), 648–660.

Getz, D. (2000) Tourist shopping villages: Development and planning strategies. In C. Ryan and S. Page (eds) *Tourism Management: Towards the New Millennium* (pp. 211–225). Oxford: Elsevier Science.

Guiry, M., Magi, A. and Lutz, R. (2006) Defining and measuring recreational shopper identity. *Journal of the Academy of Marketing Science* 34 (1), 74–83.

Jones, M.A., Reynolds, K.E. and Arnold, M.J. (2006) Hedonic and utilitarian shopping value: Investigating differential effects on retail outcomes. *Journal of Business Research* 59, 974–981.

Josiam, B., Kinley, T. and Kim, Y-K. (2005) Involvement and the tourist shopper: Using the involvement construct to segment the American tourist shopper at the mall. *Journal of Vacation Marketing* 11 (2), 135–154.

LeHew, M.L.A. and Wesley, S.C. (2007) Tourist shoppers' satisfaction with regional shopping mall experiences. *International Journal of Culture, Tourism and Hospitality Research* 1 (1), 82–96.

Machleit, K.A., Meyer, T. and Eroglu, S.A. (2005) Evaluating the nature of hassles and uplifts in the retail shopping context. *Journal of Business Research* 58 (5), 655–663.

Metcalf, B. (1995) A brief history of communal experimentation in Australia. In B. Metcalf (ed.) *Co-operative Lifestyles in Australia: From Utopian Dreaming to Communal Reality* (pp. 14–40). Sydney: University of New South Wales Press.

Mitchell, C. (2003) The heritage shopping village: Profit, preservation and production. In G. Wall (ed.) *Tourism: People, Place and Products* (pp. 151–176). Waterloo: University of Waterloo.

Mitchell, C. and de Waal, S. B. (2009) Revisiting the model of creative destruction: St. Jacobs, Ontario, a decade later. *Journal of Rural Studies* 25, 156–167.

Moscardo, G. (2004) Shopping as a destination attraction: An empirical examination of the role of shopping in tourists' destination choice process and experience. *Journal of Vacation Marketing* 10, 294–307.

Murphy, L.E., Moscardo, G., Benckendorff P. and Pearce P. (2008) Tourist shopping villages: Exploring success and failure. In A. Woodside and D. Martin (eds) *Advancing Tourism Management.* London: Elsevier.

Murphy, L., Benckendorff, P., Moscardo, G. and Pearce, P.L. (2011a) *Tourist Shopping Villages: Forms and Functions.* Oxford: Routledge.

Murphy, L., Moscardo, G., Benckendorff, P. and Pearce, P. (2011b) Evaluating tourist satisfaction with the retail experience in a typical tourist shopping village. *International Journal of Retailing and Consumer Services* 18 (4), 302–310.

Murphy, L., Pearce, P., Benckendorff, P. and Moscardo, G. (2008) Tourist shopping villages: Challenges and issues in developing regional tourism. In S. Richardson, L. Fredline, A. Patiar and M. Ternel (eds) *Tourism and Hospitality Research, Training and Practice: Where the Bloody Hell are We?* Proceedings of the 18th Annual Council of Australian University Tourism and Hospitality CAUTHE Conference, 11–14 February 2008.

Swanson, K.K. and Timothy, D.J. (2012) Souvenirs: Icons of meaning, commercialisation and commoditization. *Tourism Management* 33 (3), 489–499.

Tsiotsou, R. (2006) The role of perceived product quality and overall satisfaction on purchase intentions. *International Journal of Consumer Studies* 30 (2), 207–217.

Turley, L.W. and Milliman, R (2000) Atmospheric effects on shopping behaviour: A review of the experimental evidence. *Journal of Business Research* 49 (2), 193–211.

Underhill, P. (2004) *The Call of the Mall.* New York: Simon & Schuster.

Wilkins, H. (2011) Souvenirs: What and why we buy. *Journal of Travel Research* 50 (3), 239–247.

Yuksel, A. (2007) Tourist shopping habitat: Effects on emotions, shopping value and behaviours. *Tourism Management* 28, 58–69.

Zhuang, G., Tsang, A.S.L., Zhou, N., Li, F. and Nicholls, J.A.F. (2006) Impacts of situational factors on buying decisions in shopping malls: An empirical study with multinational data. *European Journal of Marketing* 40 (1), 17–43.

10 Souvenir Development in Peripheral Areas: Local Constraints in a Global Market

R. Geoffrey Lacher and Susan L. Slocum

The World Tourism Organization notes that tourism can make significant contributions to 'poverty alleviation, economic growth, [and] sustainable development' in rural areas that often have few alternative options to generate significant income (UNWTO, 2005: 1). Policy makers, particularly from the developing countries, frequently concur with this assessment as they develop projects to intensify the level of tourism development in their countries (UNWTO, 2007). While these projects often succeed in attracting tourists, the literature has provided numerous examples where tourism has failed to promote local development and improve the livelihoods. This failure appears to be particularly common in peripheral communities, those that tend to be rural, relatively poor and politically marginalised (Britton, 1982; Lindberg *et al.*, 1996; Mbaiwa, 2005; Nyaupane *et al.*, 2006; Walpole & Goodwin, 2000). These peripheral communities are generally ill-prepared to take advantage of the new economic opportunities afforded by tourism development. Because rural communities typically do not have the capital to create substantial businesses or the expertise to gain upper-level employment, they are often limited to low-end jobs or income. This lack of local involvement results in high external leakages that hinder economic development (Britton, 1982; Lindberg *et al.*, 1996).

Souvenirs have been suggested as an effective means to generate local economic benefits for the host community (Cohen, 1993a; Healy, 1994).

Owing to the high percentage of tourists' expenditures that goes to shopping and souvenirs, locally made handicrafts could play a key role in the economic development of the host community. Healy (1994) suggests that handicrafts offer several major advantages to residents of rural tourists destinations: they do not have to leave the rural setting to make money from tourism; handicrafts can be created during seasonal (or daily) slack times; handicrafts can provide income to women, children, the elderly and the infirm; crafts frequently require little training and capital investment; and they have low levels of external leakage. Despite their potential advantages as a means of economic development, souvenirs are still subject to the core-periphery relationship decried by critics of traditional tourism development (Britton, 1982; Mbaiwa, 2005; Walpole & Goodwin, 2000).

While some researchers have asserted that selling locally made crafts as souvenirs could help a destination avoid the negative effects of core-periphery relationships and accelerate economic development (Britton, 1982), Cohen believed that this is not necessarily the case, noting that one of souvenirs' ambiguities is whether they are 'an important supplemental source of income for people with few alternatives or a means of their exploitation?' (Cohen, 1993a: 1). The sources of souvenirs sold to tourists add to this ambiguity. While souvenirs can be locally produced, oftentimes souvenirs are mass produced in factories away from the main tourist destination. They are then sold through middlemen until they reach the local vendors at the destination who sell them to the tourists (Cohen, 1993b; Hume, 2009). With respect to economic impacts, questions arise about the degree to which higher quality, non-local souvenirs increase the demand for handicrafts. At the same time, local vendors often struggle to sell locally made crafts to tourists owing to a variety of constraints. This chapter will add to the literature by exploring the constraints that peripheral communities face when they attempt to enter the souvenir market, using data collected during recent fieldwork in Tanzania and Thailand.

Souvenirs in the Periphery

The origins of souvenirs sold to tourists are diverse and difficult to summarise (Healy, 1994); souvenirs can be made by individuals or groups, skilled craftsmen or novices, large factories or small households. Souvenirs may be produced in the same area where they are sold; however, souvenirs are also produced in factories at or away from the main tourist destination and then imported by local vendors who sell them to tourists (Hume, 2009). While there is a dearth of literature on the business of selling souvenirs in peripheral villages (some examples include Cohen, 1993 and Hume, 2009), it can be assumed that local souvenir craftsmen and vendors face many of the same constraints that peripheries face when trying to develop other tourism businesses. These constraints include a lack of capital (Forsyth, 1995), a lack of education (Fuller *et al.*, 2005), a lack of experience (Holder, 1989) and competition from outside businesses (Cohen, 1993b).

There are a number of advantages to the development of local souvenir businesses. From a cultural perspective, souvenir production can be a form of heritage conservation. Traditional designs and/or production techniques play a large part in the commercial value of souvenirs (Asplet & Cooper, 2000). Yu and Littrel (2003) argue that the art of producing souvenirs can become a tourist attraction in its own right, which encourages cultural exchange and adds value to the craft itself. Traditions that may be diluted by globalisation can be rekindled and younger generations may find new commercial outlets for their culture and history (Hume, 2009). Souvenir production can provide economic opportunities for marginalised groups – such as women, the disabled or the elderly (Healy, 1994) – if locally produced souvenirs have very low leakage and keep tourism revenue in the village (Lacher & Nepal, 2011). Additionally, handicrafts can be created during seasonal (or daily) slack times (Healy, 1994) so that they do not interfere with agricultural labour demands in these peripheral areas.

Small business development, including souvenir development, requires access to capital if profit is to be realised by the producers. There is ample evidence that rural communities in less developed countries frequently lack access to these assets of production, especially financial capital (Forsyth, 1995) and/or human capital (Fuller *et al.*, 2005; Holder, 1989; Nyaupane *et al.*, 2006) needed to run a business oriented towards tourists. This results in profits that are kept by the outsiders who own the largest firms and control the largest market share. External ownership may result in high leakage of tourism revenue and reduced opportunities for local entrepreneurs (Hampton, 1998; Milne, 1987). Peripheral areas, especially those in the developing world, generally lack the capital to run all but the most basic of services. In peripheral regions, as little as $40 may constitute a significant barrier to entering the tourism market (Forsyth, 1995). Management and marketing may also be a major problem for local ownership (Holder, 1989; Torres, 2003), as local entrepreneurs typically do not have the experience or education to market products to foreigners, and their ability to educate themselves in the matter is often limited (Holder, 1989). Given these limitations, peripheral tourism destinations frequently rely on the importation of souvenirs to meet market demands (Cohen,

1993b). Cohen (1993a: 4) summarises the problems these imported souvenirs create:

> The successful commercialization of an ethnic craft invites copycats from outside the original producers' group – whether members of other ethnic groups or even of the majority population. Their penetration of the market is facilitated by the fact that ethnic styles and products do not in most countries enjoy legal protection ... External penetration of the market has a double effect on the original producers. It not only contracts the market for their own products but frequently also leads to a fall in their prices.

In addition to competition from imported souvenirs, one would expect to see the normal lists of constraints to operating rural tourism businesses in the periphery in souvenir businesses: notably lack of financial capital (Forsyth, 1995), lack of human capital (Fuller et al., 2005; Holder, 1989; Nyaupane et al., 2006), and unfamiliarity with management and marketing (Holder, 1989; Torres, 2003). However, despite the potential relevance of souvenir production to peripheral economies, the constraints to developing souvenir production and businesses in rural tourism destinations are not well understood or documented. This chapter attempts to fill this gap by providing case studies from Northern Thailand and Tanzania that illustrate the potential challenges in reducing leakages from rural areas.

Study Area

Northern Thailand

With the exception of Chiang Mai (population 700,000), Northern Thailand is primarily a rural region dotted with small towns and market centres. The mountainous nature of the region provides many opportunities for the tourists to enjoy trekking, rafting, jungle safaris and encounters with hill-tribe communities. 'Hill tribe' is a name given to an assortment of approximately 20 culturally distinct ethnic groups that inhabit the highlands of Southeast Asia. Six ethnic groups, which include the Karen, Hmong (Meo), Mien (Yao), Lahu, Lisu and Akha, have been the centre of attraction for 'hill tribe tourism'. Because of the hill tribes' linguistic and cultural differences, as well as their inexperience with the modern economy, they remain on the margins of Thai society. The village that was most reliant on souvenirs for income was Huay Pu Keng, one of three villages populated by the Kayan, a subset of the Karen. Many of the Kayan women wear brass coils around their neck, which deform their shoulders to appear as if their necks are elongated. This custom makes Huay Pu Keng a very popular tourism destination. The Kayan were displaced from the Karenni state in central eastern Myanmar, across the border from Mae Hong Son Province. They are considered by the Thai authorities to be 'refugees' or 'chon gloom noi' meaning their movement is highly restricted by the authorities. Another village investigated was, Pong Ngean, a Karen village of about 50 people located 100 km north of Chiang Mai. The village was established in the early 1980s mainly to host tourists. Its main attractions include the hill-tribe culture and bamboo rafting. Mae Aw, a Kuomintang (KMT) village two hours north of Mae Hong Son, was also interviewed. At one time the KMT were part of the Chinese Nationalist political party and military force led by Chiang Kai-shek. Because of their Chinese heritage, the villagers remain culturally distinct from the Thais and their village is a popular tourism destination for domestic and international tourists. The final village, Tom Lod, is inhabited by those of the Shan ethnicity, a subset of the Thai ethnic group that is common in Northern Thailand. Its main attraction is a large cave and a river running through it.

Northern Tanzania

Tanzania has positioned itself as Africa's leading upscale safari destination. Northern Tanzania has become a well-known tourism destination owing to the abundant wildlife and cultural diversity of the area. Because a large portion of land is protected around Arusha National Park, the inhabited areas are some of the most densely populated areas in Tanzania. The region is divided into five districts, of which the first research site, Arumeru District, houses just above 40% of the region's population. Arumeru is located directly between the urban centre of Arusha Town and the Mount Kilimanjaro International Airport, and encompasses Mount Meru and the entrance to Arusha National Park. The rich natural amenities in Arumeru District have encouraged a number of five-star hotels and resorts throughout the district. Arumeru District is a series of tightly packed urban villages with many of the same social constructions and agricultural-based industries as the rest of Tanzania; however, the sheer volume of inward migration leads to urban problems, such as overcrowding, unemployment and crime. All tourists that visit the northern tourist circuit must pass through Arumeru District. In total, five villages and one woman's group participated in the focus groups. These villages include: Maji Ya Chai, Nguruma, Patandi, Uswahilini, Ngongongare and the woman's group called the Usa River Widows Group and was comprised of HIV widows.

The second research site, Pangani District, is located in Tanga Region, 60 miles south of the City of Tanga along the Indian Ocean. The people of Pangani District are known as the Swahili people and their language was adopted as the official Tanzanian language. They believe they are direct descendants of the lucrative Arab slave trade and most residents are of the Muslim faith. Pangani is the northern hub to Saadani National Park, and most tourists travelling overland to the park must pass through the area. Pangani is also home to ancient Arab heritage, unique historical architecture (both Arab and German) and spectacular beaches that rival Zanzibar. The area is relatively new to tourism, although tourist populations are increasing, along with the development of tourism infrastructure. The two major destinations are Pangani Town and Ushongo, a village located approximately 30 miles south of Pangani Town. Ushongo has an air strip that can accommodate small planes, and there is an airstrip and a golf course planned for Pangani Town. There is no paved road south of Tanga City and the heavily trafficked road suffers during the rainy seasons. Pangani District has encountered a huge population increase that many locals feel is associated with the increased employment opportunities through tourism expansions in the area. There were six villages that participated in focus groups including: Saange, Ushongo, Matikani, Sakura, Bushiri and Kipumbwi.

Methods

This research is based on recent fieldwork by the authors on the tourism economy in villages in Tanzania and Northern Thailand. By incorporating such diverse locations, limitations can be compared so as to determine which constraints are cross-cultural. While these projects were done independently by different researchers, the methodologies were similar, both following qualitative methodologies based on understanding limitations to participation in the tourism industry. Given the exploratory nature and interpretivists aims of the projects, a qualitative methodology was appropriate in both cases (Fadahunsi, 2000). In total 12 villages in Tanzania and four villages in Thailand were investigated and 38 qualitative interviews were conducted and transcribed, with the assistance of translators. These interviews focused on a wide variety of topics including the history of local tourism, the constraints faced by locals attempting to become involved in the tourism industry and their hopes for the future of tourism as an economic development tool in the area. Souvenirs were common topics of discussion, as they are a key source of income in several of the villages and villagers often voiced their frustrations with the difficulties they faced when trying to become involved in the souvenir industry.

A brief synopsis of the fieldwork for both projects is included here. A complete overview of the two projects in their entirety can be found in Lacher and Nepal (2010) and Slocum *et al.* (2011).

Interviews in Thailand were done individually, with 26 informal interviews with local residents across four different villages in the Chiang Mai and Mae Hong Son provinces. The major stakeholders in the tourism industry were selected. The interviews averaged 30 minutes in length. Many respondents were contacted more than once to seek clarification regarding their previous responses. Interviews in the villages focused on understanding the history of tourism in the area, local complaints and obstacles to participation in the tourism industry, strategies to mitigate challenges and future outlook of the industry. All interviews were transcribed in English in the field by and with help from the translators.

In Tanzania, focus group interviews were conducted in 12 villages within the Arumeru District, near Arusha. Participation rates ranged from six to 15 people. The sampling technique for participant villages in Arumeru District utilised snowball sampling. The initial village was approached based on a recommendation from a hotel establishment that had begun community development programmes in the area. From there, a second village was recommended by the first village's administrator and so on. In Pangani, purposive sampling was utilised and villages were selected based on their exposure to tourist and tourism infrastructure. For example, villages that had hotel accommodations, established village tour operations or major transportation hubs were used. Some focus groups were to be broken into categories as a means to recognise 'the importance of constituting groups in ways that mitigate against alienation, create solidarity, and enhance community building' as a means to create homogeneous groups (Denzin & Lincoln, 2005: 896). An all-women's group, a tour-guide group and a business-ownership group were conducted, as well as a number of mixed community groups.

Once specific case studies were independently derived, a cross comparison of the themes developed offered insight into reoccurring or region-specific constraints. Yin (2003) defines 'analytical generalization' when 'a previously developed theory is used as a template with which to compare the empirical results of the case study' (Yin, 2003: 32–33). Using an iterative process, the strength of this methodology lies in the inductive approach that provides suggestive rather than definitive analysis (Welch, 1994). Using cross-case comparisons, emergent patterns provide similarities and differences between the two research sites and 'while this methodology requires extensive study into the lived experiences of villagers in each site, the depth and breadth of knowledge resulting from this methodology allow for a more rigorous interpretation and understanding of economic constraints' (Slocum *et al.*, 2012: 537). For data analyses, quotes relating to the constraints faced by locals were extracted from the interviews, and then similar quotes were then sorted into themes identified by the researchers. Themes were developed by inductively sorting the extracted items. In the results section these themes are described and relevant examples are provided.

Results and Discussion

Six themes involving constraints were found. These include: (1) lack of marketing skills; (2) lack of knowledge of tourists' demand for souvenirs; (3) lack of capital; (4) lack of access to tourism markets; (5) competition from outside areas; and (6) natural resource depletion (see Table 10.1). These themes are briefly defined, and illustrated through an observed statement or example.

Table 10.1. Constraint comparison between Thailand and Tanzania

Theme	Tanzania	Thailand
Lack of marketing skills	• Tour guides control tourism shopping locations • Ethnic and religious entry barriers	• No 'locally made' labelling • Craft distribution limited to immediate villages
Lack of knowledge of tourists' demand for souvenirs	• Lack of knowledge of consumer preferences	• Lack of knowledge of consumer preferences
Lack of capital	• Lack of start-up capital • No credit history • Lack of discretionary time outside farming	• Lack of start-up capital • Fear of investing • Lack of discretionary time outside farming
Lack of access to tourism markets	• Low tourist numbers • Poor infrastructure • High commission charges	• Low tourist numbers • Political isolation • Poor infrastructure
Competition from outside areas	• International imports of souvenirs • Tourists geared towards stereotypical African art	• Urban imports of souvenirs • Fear of adequate quality
Natural resource depletion	• Tourists wary of protected natural inputs	• Regulation of bamboo harvesting

Lack of marketing skills

This research found that peripheral villagers do not understand how to market goods to western tourists. Nowhere in the Thai villages were goods advertised specifically as locally made, even though tourists' desire locally made products, and the specialty souvenirs of the villagers are not marketed to tourists outside of the villages. While the Longneck women of Huay Pu Keng were expert weavers, crafting a variety of uniquely styled scarves, their crafts are not marketed to tourists outside the village; rather their unique cultural practices are used to draw tourists into the village. In Tanzania, villagers complained that they were unable to sell souvenirs because most tourists were directed to other established dealers with better advertising as tourist guides brought tourists to businesses willing to pay large commissions. Additionally, Tanzanian culture favours family or ethnic ties resulting in entry barriers to tourism markets by marginalised tribal groups with little political or social clout. In Pangani, the influx of educated Christian immigrants from rural areas has resulted in new supply chains that import crafts from outside the area, particularly from the Masaai villages in the north and the island of Zanzibar. In Arumeru, the long-established urban ties exclude rural immigrants looking for new employment and business opportunities.

Lack of knowledge of tourists' demand for souvenirs

Peripheral villagers may feel helpless as they do not know what types of souvenirs tourists' desire. In both Tanzanian and Thai villages, the locals often asked the researchers for advice on what westerners want, and would bemoan their lack of knowledge of western tastes. Entrepreneurs largely employed a copy-cat method of souvenir production, emulating the souvenirs they saw sold in other tourism areas. One Thai villager was surprised when a tourist asked about whether a small decorative piece in his house (rather than in his store) was for sale. While he had not previously considered selling this item, he eagerly negotiated a price with the tourists and began manufacturing them to include in his store. Other villagers showed themselves to be experts at making crafts out of bamboo, but did not believe that these crafts would be of interests to tourists. In Tom Lod, the locals have largely ignored the potential of souvenirs and allow Lahu and Lisu craftsmen to come into the village and sell their wares. The Shan people in Tom Lod feel largely inexperienced about souvenirs in general, not understanding what they could produce that tourists would want. In Tanzania, villagers emulate the successful Masaai bead work, flooding the market with lower-quality reproductions. Furthermore, reed mats, traditionally used as prayer mats, drew little attention from western travellers. It was suggested that these reed construction techniques could be used to make useful household items for westerners such as table mats or coasters, and potentially create new types of souvenirs for sale.

Lack of capital

Peripheral villagers may not be able to afford the tools and materials needed to begin a souvenir manufacturing business. Loans were very difficult to obtain in Thailand. To initiate a new business, an individual would be largely reliant on family members to provide the necessary capital to start a business. This is compounded because villagers are afraid to invest their meagre resources into an industry that they do not fully comprehend. Villagers in both countries needed assistance from Non-Governmental Organisations or government organisations, in the form of increased access to micro-finance in order to raise the money necessary to begin souvenir production, but they lack knowledge on how best to solicit this help. Additionally, while micro-financing was beginning to appear in Tanzania, rural populations had no convertible assets or repayment history, leaving them disadvantaged over more urban populations. While beginning a souvenir business may require minimal capital investment, it may require a great deal of time. Individuals in these villages have responsibilities that keep them tied to the farm plot, or may not have the savings needed to allow them to go days without wages as they risk a new career in souvenir sales. Villagers frequently cited uncertainty when discussing their hesitation to be involved with the tourism industry.

Lack of access to tourism markets

Peripheral villagers often find it difficult to get their souvenirs seen by tourists. Some villages did not have enough tourists to support their own souvenir market. The cost and difficulty in transporting these items makes entering new markets challenging. Other times, the villagers struggled to convince luxury hotels to carry their locally produced souvenirs in gift shops. The Kayan in Huay Pu Keng were not typically allowed into the larger cities and tourist-oriented markets owing to their status as refugees. Other villagers were often unable to gain access to larger markets owing to the high cost of travel, or in some cases the total lack of transportation. The tourism traffic in the larger destinations, such as Chiang Mai or Mae Hong Son, is many times that of even the most popular rural villages, but the villagers are largely unable to access these larger markets. In Tanzania, poorly developed roads further created access difficulties and added to the sporadic supply of indigenous crafts. Large commissions were regularly paid by larger craft shops to encourage guides to include them as a stop on an organised tour. Lastly, tourists rarely walked unaided through rural villages, preferring a driver or guide to escort them on their outings.

Competition from outside areas

Goods from outside, frequently more urban areas, often come into competition with goods locally made in peripheral villages. A survey of businesses in a village in Thailand showed that 76% of souvenirs sold were imported from urban areas; the villagers were concerned that the goods they made were not as high quality as the imported goods, and thus decided against competing against the imported souvenirs with crafts of their own. In Arumeru, Chinese imports have recently flooded the market with copies of traditional African carvings and paintings. Furthermore, this influx of stereotypically African art has led tourists away from the more traditional weaving and pottery souvenirs that the residents have knowledge in producing. While the globalisation of tourism has given these individuals the opportunity to profit from tourism, they also encounter competition from core areas. Urban areas may have the advantages of having more widely known areas from which to purchase souvenirs. Guidebooks, such as the Lonely Planet, point tourists towards markets and bazaars in the more urban areas, but offer little in discussing the locally made crafts that can be acquired in the rural villages.

Natural resource depletion

Tourism is often concentrated in protected areas that limit access to traditional craft inputs. Raw materials, such as natural grasses, certain species of trees (i.e. teak and ebony), coral and shells, are often conserved as part of the tourist resource and local populations may have to pay to import these items. In Tanzania, conservation initiatives have reduced the availability of natural craft inputs and raw materials have become heavily regulated. Ebony, ivory, aquatic species and other inputs are not only harder to source, but tourists are often aware of conservation efforts to protect these species and shy away from buying them. Furthermore, conservation organisations have introduced new natural species for locals to use as source material, but these have so far failed to entice tourists to buy as the crafts appear different from what has traditionally been produced. Despite the plants resilience, bamboo harvesting is highly regulated in some parts of Thailand, at least partly owing to its use in constructing rafts on which to ferry tourists.

Conclusion

While souvenirs appear to be an ideal means of augmenting economic development in peripheral villages with a tourism industry, a number of constraints frequently keep locals from taking full advantage of the opportunities that souvenirs provide. Souvenirs can play a major role in the economics of peripheral villages but appear to be almost entirely neglected. A complex array of opportunities, constraints, tourist types, artisanal skills and local ascetics determine the role that souvenirs play in the local economy. Healy (1994) and Britton's (1982) assentation that souvenirs may represent an ideal means of supplementing local economic development may be correct, but it remains a complicated path.

This research would suggest three potential solutions to these issues: clearly identifying locally made goods, forming cooperatives and providing organised assistance. By identifying locally made goods and thus distinguishing them from imported goods, local villages may increase the amount of local goods they sell and thus reduce leakages. The villagers should also ask the tour guides to emphasise the uniqueness of the souvenirs to encourage their purchase instead of outside imports. This may persuade many consumers to purchase the locally made souvenirs because of their authenticity and uniqueness, especially the segment of tourists that Littrell *et al.* (1993) identify as specifically seeking authenticity. Diverting purchases away from imported souvenirs to local souvenirs should increase the sellers' income by reducing leakages. While this might be a simple step, many souvenir vendors in the developing world, particularly at remote tourist destinations like Huay Pu Keng and Pangani, may not fully grasp the desires and demands of western tourists or lack the language and advertising skills to express which souvenirs are locally made and which ones are imported. This strategy would be fairly easy to implement, but may face resistance from vendors that sell mainly imported goods.

New cooperatives should provide a number of positive influences on tourism production. Members can work to ensure that resources are conserved, locally made items are prominently displayed, social capital and borrowing power are increased for local groups and create a more reliable supply of crafts to local markets. The USA River Widow's group has been very successful in promoting their wares by advertising that proceeds go to widows of HIV and, through the cooperative, have managed to keep local gift shops and hotels well stocked with quality merchandise. Additionally, cross-village cooperatives can be created. These cooperatives can promote locally made crafts to the larger population, and provide access to more resources and knowledge about promoting items to western tourists.

Finally, some form of organised assistance from locally based NGOs, universities or government offices could be provided to villages to assist with their marketing efforts and technical skills. Education, both in basic business skills and in the understanding of tourism markets, helps local producers adapt and service foreign consumers. Education also helps local groups write basic business plans that can be used to secure micro-financing and other forms of financial capital. By promoting the unique qualities of their goods, such as the use of more sustainable resources or the traditional uses of local crafts, tourists can be encouraged to buy locally produced goods rather than imported goods. This assistance might also include financial assistance such as grants or loans that could help potential entrepreneurs start their businesses.

Navigating the opportunities and threats presented by the globalised economy and international tourists can be a disorienting experience for peripheral villages in the developing world. Hard work and intelligent strategies are needed to help negotiate the constraints faced by potential craftsmen and entrepreneurs and ensure that these villages benefit from the process of globalisation. Despite the noted constraints, souvenirs appear to be a successful way of moving peripheral economies forward. Even if the impact is somewhat limited, souvenirs do contribute to the local economy (see Lacher and Nepal (2011) for details), and villagers are interested and eager to expand souvenir production and sales.

Acknowledgement

The authors wish to acknowledge that funding support for the research on which this paper is based was provided by the United States Department of Agriculture (Agreement # 58-3148-5-149).

References

Asplet, M. and Cooper, M. (2000) Cultural designs in New Zealand souvenir clothing: The question of authenticity. *Tourism Management* 21 (3), 307–312.

Britton, S. (1982) The political economy of tourism in the third world. *Annals of Tourism Research* 9 (3), 331–368.

Cohen, E. (1993a) Introduction: Investigating tourist arts. *Annals of Tourism Research* 20 (1), 1–8.

Cohen, E. (1993b) The heterogenization of a tourist art. *Annals of Tourism Research* 20 (1), 138–163.

Denzin, N. and Lincoln, Y. (2005) *The Sage Handbook of Qualitative Research* (3rd edn). Thousand Oaks, CA: Sage.

Fadahunsi, A. (2000) Researching informal entrepreneurship in Sub-Saharan Africa: A note on field methodology. *Journal of Developmental Entrepreneurship* 5 (3), 249–260.

Forsyth, T.J. (1995) Tourism and agricultural development in Thailand. *Annals of Tourism Research* 22 (4), 877–900.

Fuller, D., Buultjens, J. and Cummings, E. (2005) Ecotourism and indigenous micro-enterprise formation in northern Australia opportunities and constraints. *Tourism Management* 26 (6), 891–904.

Hampton, M.P. (1998) Backpacker tourism and economic development. *Annals of Tourism Research* 25(3), 639–660.

Healy, R.G. (1994) Tourist merchandise as a means of generating local benefits from ecotourism. *Journal of Sustainable Tourism* 2 (3), 137–151.

Holder, J. (1989) Tourism and the future of Caribbean handicraft. *Tourism Management* 10 (4), 310–314.

Hume, D.L. (2009) The development of tourist art and souvenirs – the arc of the boomerang: from hunting, fighting and ceremony to tourist souvenir. *International Journal of Tourism Research* 11 (1), 55–70.

Lacher, R. and Nepal, S. (2010) Dependency and development: Tourism in Northern Thailand. *Annals of Tourism Research* 37 (4), 947–968.

Lacher, R.G. and Nepal, S.K. (2011) The Economic impact of souvenir sales in peripheral areas: A case study from northern Thailand. *Tourism Recreation Research* 36 (1), 27–38.

Littrell, M.A., Anderson, L. and Brown, P.J. (1993) What makes a craft souvenir authentic? *Annals of Tourism Research* 20 (1), 197–215.

Lindberg, K., Enriquez, J. and Sproule, K. (1996) Ecotourism questioned: Case studies from Belize. *Annals of Tourism Research* 23 (3), 543–562.

Mbaiwa, J. (2005) The problems and prospects of sustainable tourism development in the Okavango Delta, Botswana. *Journal of Sustainable Tourism* 13 (3), 203–227.

Milne, S.S. (1987) Differential multipliers. *Annals of Tourism Research* 14 (4), 499–515.

Slocum, S.L., Backman, K.F. and Robinson, K.L. (2011) Tourism pathways to prosperity: Perspectives on the informal economy in Tanzania. *Tourism Analysis* 16 (1), 43–55.

Slocum, S.L., Backman, K. and Baldwin, E. (2012) Independent instrumental case studies: Allowing for the autonomy of cultural, social, and business networks in Tanzania. In K. Hyde, C. Ryan and A. Woodside (eds) *Field Guide for Case Study Research in Tourism*. Bingley: Emerald Publishers.

Torres, R. (2003) Linkages between tourism and agriculture in Mexico. *Annals of Tourism Research* 30 (3), 546–566.

Walpole, M.J. and Goodwin, H.J. (2000) Local economic impacts of dragon tourism in Indonesia. *Annals of Tourism Research* 27 (3), 559–576.

UNWTO (2005) *Declaration – Harnessing Tourism for the Millennium Development Goals.* Accessed 4 August 2009, http://www.unwto.org/step/pub/en/pdf/declaration.pdf

UNWTO (2007) *ST-EP Programme.* Accessed 4 August 2009, http://www.unwto.org/step/pub/en/pdf/step_prog.pdf

Yin, R.K. (2003) *Case Study Research: Design and Methods* (3rd edn). Thousand Oaks, CA: Sage.

11 Souvenir Production and Attraction: Vietnam's Traditional Handicraft Villages

Huong T. Bui and Lee Jolliffe

> *Wander our narrow streets and see the vendors with their art ware, tableware and ornaments. Little animals, figurines, huge vases, tea sets. Studios with antique collections, craftspeople making the pots, applying the intricate designs, setting the kilns. Packers loading trucks for all parts. Walk to the river, see the old kilns and look for their red fires. Enjoy a tea, it tastes better in Bat Trang! Buy some ceramics so you can enjoy your Bat Trang experience at home.*
>
> CK&T Ceramics, 2006

In Vietnam, traditional arts and crafts play an important role in souvenir production. Many crafts are made on-site in handicraft villages. Handicraft villages such as Bat Trang described above, typically located on the peripheries of city and town destinations, usually focus on one type of traditional craft. These crafts are purchased by tourists as souvenirs, either at the villages or in retail souvenir outlets in nearby towns and cities.

Vietnam is an emerging economy with a significant international and domestic tourism market, receiving 6 million international tourists in 2011 (Viet Nam Administration of Tourism, 2012). Since 1986, the country has undergone radical economic and international relations reform. Vietnam's population is 89 million people. One in three residents travelled within the country in 2009 totalling 25 million domestic travellers (Viet Nam Administration of Tourism, 2010). The Master Plan of Vietnam Tourism, 2008–2020 (Viet Nam Administration of Tourism, 2005) emphasised visits to traditional handicraft villages in order to diversify tourism products and services.

In the context of diversification through handicraft production, this chapter addresses the central research question of how handicraft villages on the peripheries of cities and towns in Vietnam are transitioning from production to attraction in terms of the creation of place-based souvenirs. The chapter focuses on case studies profiling two handicraft villages, one on the periphery of a historic city, the other on the edge of an ancient town.

Literature Review

Vietnam's handicrafts have developed over hundreds of years, with more than 200 different types of products now produced (including pottery, paper, ceramics, earthenware, rattan and bamboo and silk). Traditionally, agricultural workers made handicrafts for their own needs after the harvest, over time this led to the production of village-specific specialties in particular types of crafts. Vietnam's Ministry of Agriculture and Rural Development has promoted handicraft production for export as a way to transition from an agriculture-based economy to a manufacturing and service economy. Since handicraft production affects local livelihoods, a number of countries have assisted Vietnam to improve handicraft production for export. For example, The Japan International Cooperation Agency completed a Study on Artisan Craft Development Plan for Rural Industrialisation of Vietnam (Japan International Cooperation Agent, 2004) and the German Technical Cooperation completed a study on craft villages in Quang Nam (German Technical Cooperation, 2007).

Traditional handicraft villages are located in rural areas, with at least 30% of households participating in producing hand-made products based on traditional knowledge (Ministry of Agriculture and Rural Development, 2002). The Vietnam Handicraft Research and Promotion Centre notes there are almost 1500 handicraft villages in Vietnam, with a high density in Northern Vietnam (Luong, 2006). About 300 of these are traditional craft villages. Handicraft development in Vietnam can be divided into three periods (Table 11.1).

The 1986 government policy of 'doi moi' (Vasavakul, 2001) put in place a move from a socialised to a market economy in Vietnam. This meant that large numbers of handicraft villages transitioned from production for domestic markets to export markets and at the same time developed a souvenir focus. Handicraft production as entrepreneurial diversification can be seen as a growing trend in Vietnamese villages, spreading from urban to rural villages, responding to the shift in the economic platform (Bryant, 1998). Vietnam is located, in terms of leisure, in the peripheral area of relatively poor tourist receiving countries identified as 'less developed countries' (Weaver, 1998).

Table 11.1 Handicraft development in Vietnam

Period	Characteristics
Prior to the 20th century	Handicrafts made for domestic (home) and farming use
During the 20th century	Scope of handicraft production widened for export
Since the end of the 20th century	Handicrafts mainly produced for export and tourism souvenirs

Source: Japan International Cooperation Agency (2004)

The periphery has long been an important concept in tourism study (Turner & Ash, 1975; Weaver, 1998). Peripheral areas are characterised by a number of factors; (1) geographically remote from mass market (Hall, 2005), (2) lack effective political and economic over decisions affecting their well-being (Hall & Jenkins, 1998); (3) internal economic linkages tend to be weaker at the periphery than at the core (Archer, 1989); (4) migration flows tend to be from peripheral to the core (Hall, 2007), (5) the national and local states play more significant interventionist roles (Hall & Jenkins, 1998); (6) peripheral regions often retain high aesthetic amenity values (Hall & Muller, 2004). Thus, Hall (2007) states governments at various levels are seeking to find means to support peripheral regions, within which tourism is often seen as a vehicle for regional development.

Tourism development is often viewed as a manifestation of the 'global–local nexus' (Chang et al., 1996), particularly in relation to local cultural heritage (Teo & Lim, 2003). According to Szydlowski (2008: 59) 'globalization and tourism have introduced western consumers to Vietnamese handicrafts, which are attractive to tourists because they are able to connect the product to the village where it was created'. Souvenirs are produced by small-scale producers, working independently or through co-operatives, or by larger-scale mass production units (Swanson & Timothy, 2012). Producers may sell souvenirs directly to tourists or distribute through intermediaries for sale by retailers elsewhere (Cohen, 1995; Rutten, 1990).

Demand for souvenirs from tourists can indirectly promote local business development and commercialisation of products. In Indonesia, Dahles (2010) identified the role of small local enterprise in tourism production that contributes to the local economy. In Thailand, Cohen (1995) identified how local crafts over time are promoted, not only through direct sales, but increasingly sold to foreign exporters through intermediaries. In the Philippines, Rutten (1990) studied the commercialisation of craft villages that produce giftware, drawing attention to the significant contribution made by the process of rural commercialisation. In Vietnam, Nguyen (2009) identified a parallel transition of traditional craft villages to commercial clusters. Souvenir production contributes to preserving traditional production techniques while sustaining the development of local businesses in developing economies (Hampton, 2005).

Local communities can increase income from selling souvenirs directly to tourists, as souvenir shopping is a popular activity (Timothy, 2005; Swanson & Timothy, 2012). Tourist shopping villages are known to appeal to tourists (Getz, 1993). Tourist shopping in handicraft villages can have local economic benefits (Healy, 1994). According to Cohen (1995), research has focused either on studies of tourist shopping behaviour (Littrell, 1990) or on the location, distribution and structure of tourist shopping facilities (Getz, 1993). The latter area has been less studied, particularly in a developing world context. Cohen's (1995) work on touristic craft ribbon development in Thailand addressed this literature gap, examining how craft-producing villages evolve. The author observed that as villages develop, rather than selling locally produced crafts, they gradually specialise in the retailing of crafts produced elsewhere and spread out on the village peripheries. Eventually, some 'craft ribbons' (streets or highways with concentrations of craft shops primarily or exclusively selling crafts, craft workshops and other service related shops) develop into tourist attractions in their own right (Cohen, 1995).

Handicraft villages featuring small-scale household production can be considered as a type of tourist shopping village, which can on occasion transition to a destination attraction (Moscardo, 2004). Tourists look for local crafts to purchase at a reasonable price (Littrell et al., 1994). Handicrafts purchased from crafts persons or in craft-production settings, such as at handicraft villages, have the advantage over those purchased elsewhere, of being perceived as authentic (Littrell et al., 1993). Meeting traditional crafters and artisans, tourists' form a connection with producers, increasing the inherent value of the item to them. Wilkins (2011) found that tourists preferred purchasing items with a place-specific regional focus, rather than more general items. Handicrafts, because they are objects made by hand using traditional knowledge, fulfil both the tourists' desire for authenticity and for a locally produced souvenir (Fairhurst et al., 2007). This chapter addresses, in part, what Cohen (1995) identified as a neglected area of study, the systematic study of tourist shopping facilities in developing countries.

Methodology

Yin (2009) indicates that the case study is an empirical inquiry investigating a contemporary phenomenon indepth and within its real-life context. To investigate the experiences of the peripheral Vietnamese handicraft

Table 11.2 Case study selection criteria

Criteria	Bat Trang	Kim Bong
Geography	Northern Vietnam	Central Vietnam
Proximity to tourist generating area	Hanoi, city	Hoi An, town
Tour services	Tours to workshops to make products	Tours to observe household production
Souvenir outlets	Craft Market	Craft Centre
Involvement in tourism	Since the 1990s	Since 2000

villages as they transition from production to attraction, a comparative case study method was therefore chosen. Case studies typically are based on a range of data sources, Yin (2009) identifies six: documentation, archival records, interviews, direct observation, participant observation and physical artefacts. In developing the comparative case studies of handicraft villages, the researchers used all of these sources, drawing on existing documentation and records on the handicraft villages, five on-site interviews with local officials in one of the case villages (Kim Bong) and directed and participant observation through two site visits to the other case village (Bat Trang), as well as examining the handicrafts produced at both villages. At Kim Bong on-site interviews were conducted in 2007. At Bat Trang, site visits were made in 2005 and 2007. The selection of the handicraft villages of Bat Trang (ceramics) and Kim Bong Village (carpentry) is a convenience sample, also justified by a number of criteria (Table 11.2).

Bat Trang

The village of Bat Trang is located in Gia Lam District, Hanoi on the Hong River 7 km downriver, or 10 km by motorway, from the city (Figure 11.1). The art of Bat Trang ceramics date from the 15th century, with artistic excellence having peaked in the 18th and early 19th centuries (Goodman, 2005). Today, the village is thriving as a producer of porcelain and ceramics, both useful and decorative, produced in Bat Trang on a large scale, mostly for domestic use or export.

Bat Trang Commune has a population of 7528 (1721 households). The majority of villagers are crafts persons, 84% of the working-age population is involved in both ceramics and porcelain production (Giang, 2009). According to Sarkun and Banerjee (2007), over 1000 families here produce 50 million assorted ceramic products every year. Yet others are employed

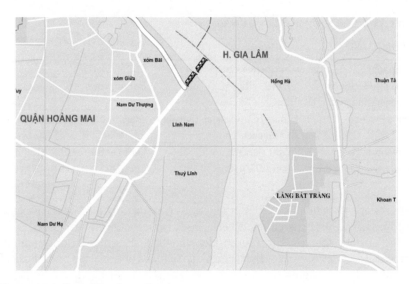

Figure 11.1 Map of Bat Trang Commune

supporting this industry, for example as observed during fieldwork, supplying slip, clay and enamel and promoting the craft.

Despite the emphasis on production, tourism has potential to contribute to livelihoods here, by increases in souvenir sales. Each year the village receives around 70,000 international and domestic visitors (Dang, 2008). Some participatory tours exist where tourists can make and decorate ceramics. The potential for tourism is noteworthy in view of recent global economic crises, as there is some evidence (Giang, 2009) the crisis has impacted sales for the village, therefore additional tourist revenues would be welcomed. However, the level of local economic impact from tourism is likely to be limited to a low level of souvenir sales, as there are neither accommodation facilities nor restaurants geared to tourists.

There is no entry fee to the village, which can be considered an un-gated industrial tourism attraction. The appeal of the village is as a site of production in a historic setting that has existed for centuries. Heritage resources include a local temple and pagoda. The village has not been contrived for tourism, but is a working production centre as well as an authentic village where people live and work. Strolling the village visitors can witness the various stages of the ceramics creation process.

The close proximity to Hanoi means that Bat Trang is on the tour itineraries for both craft village and general tours. Full-day craft village tours often combine visits to Van Phuc (silk producer) and Bat Trang (ceramics) villages.

Bat Trang is accessible for independent travellers by the Long Bien–Bat Trang public bus, a half hour ride from central Hanoi. A self-guided tour brochure is available at outlets in Hanoi in English. The village could benefit from cruise tours along the Hong River, however a wharf proposed in 2009 to allow visitation by water has not materialised.

There is evidence that Bat Trang craftspeople have the ability to innovate in terms of their adoption of information and communication technologies for their ceramics businesses. According to Tran (2003), about 15% of the ceramics producing households adopted the internet to connect with buyers. From informal discussion during a 2005 field visit to the Bat Trang market, the authors determined that several market sellers would welcome training to better equip them to serve their international tourist customers. Informal discussions with village souvenir sellers revealed that new tourism products have been introduced in Bat Trang. Since the late 2000s, the village has introduced both participative tourism services allowing tourists to make souvenirs, and educational tours for school children to learn traditional handicraft methods. Bat Trang retains manufacturing home use or decorative purposes for domestic clientele. The centre of the visitation is the Ceramic Market, where both tourism souvenirs and homeware ceramics can be found.

Various stakeholders participated in the transition of the village towards tourism, including governmental and private sectors in destination marketing. National tourism marketing strategy recognises crafts villages as an important element of the country's cultural attractions (Viet Nam Administration of Tourism, 2004). Bat Trang, thus, was included in the campaign to promote handicraft villages for the 2010 celebration of the 1000th anniversary of the establishment of Thang Long – Hanoi (Friends of Vietnam Heritage, 2010). The local private sectors play a key role in turning this traditional ceramic manufacturing village into a tourist destination. Local people initiated visitor tours and have varied their souvenir supply. The dynamics of, and trend for, innovation by locals has been attributed to their long history producing ceramics. Tourist facilities, such as ceramic workshops and souvenir shopping sites, have been initiated by local businesses. Tourism operators from Hanoi have also contributed significantly, recognising the contribution that the village makes to the Hanoi 'tourist map'.

However, tourism in the village is facing challenges to sustain its development. Local ceramic manufacturers admitted that Bat Trang is facing fierce competition in terms of price and variety from tourism souvenirs imported from China. Dang (2008) reported that about 15% of products are imported from China and the products of the village have no trademark. Another challenge is environmental pollution. The traditional method of using coal discharged air pollution and remains from the workshops were dumped to the river, creating water pollution (Dang, 2008).

Kim Bong

The village of Kim Bong is located in Quang Nam Province of Central Vietnam (Figure 11.2), a 15-minute boat trip on Thu Bon River from the UNESCO World Heritage property, Hoi An Ancient Town. The village has a history of making wooden items, dating back to the 17th century. Village carpenters from Kim Bong were believed to carve sophisticated wooden works in the famous ancient Hoi An Town and Hue Citadel (Friend of Vietnam Heritage, 2006). Village workers continue this tradition, renovating and maintaining the old wooden houses in Hoi An. The main orders for village products are for housing and architectural items, wooden ships and weaving mats (Hoi An Tourist Information Centre, 2007).

The village has now diversified its production of artistic wooden items into the manufacture of hand-made wooden tourism souvenirs (International Trade Centre, 2005). The village has also introduced new products, such as lanterns and bamboo mats, in addition to its traditional wooden articles. Gradually, the village has become a main supplier of tourism souvenirs for Hoi An Town and nearby areas. This diverse range of

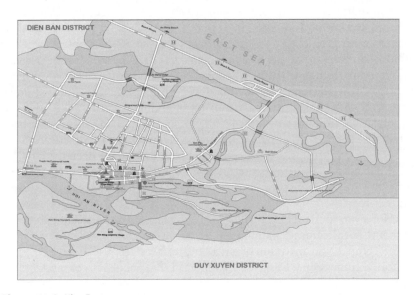

Figure 11.2 Kim Bong map

souvenir products, versus the single item focus of many of the other handicraft villages in Vietnam, is a unique characteristic of this village (German Technical Cooperation, 2007).

Recognising the benefit of the tourism souvenir business to the local economy, the village set up a Craft Centre near the wharf where tourists arrive by boats and then buy their tickets. The Craft Centre functions as an interpretation centre where visitors are introduced to the history of wood carving here and learn about the range of local hand-made wooden products via the exhibits and displays. There is also retail counter where tourists can buy souvenirs.

According to the Craft Centre Director, the traditional handicraft business of the village faces a number of challenges. First, local products have to compete against imported items from China, thus in the Craft Centre, local products account for only approximately half of the displays (SNV, 2007). The imported souvenirs are cheaper and of better variety than the local ones, thus being more attractive to the tourists. Second, the village is facing a labour shortage, as the younger generation prefers to migrate to work in urban centres for better income and a wider choice of jobs rather than continuing with their family handicraft production tradition.

Kim Bong has developed a connection to the tourism industry through its souvenir supply and a range of new services that relate to production. The village has the potential for countryside tourism, offering outdoor activities such as boating and cycling, owing to its rural setting. Owing to location, the village benefits from visitation to Hoi An. On average, the village receives approximately 30,000 international and domestic visitors each year, or about 40–100 visitor per day. The high season is from January to March with about 3900–4300 visitors (SNV, 2007). Most visitors are international tourists taking a boat day-tour and spending about 10–20 USD on souvenirs at the villages around Hoi An that include Kim Bong (German Technical Cooperation, 2007).

The village cooperative provides bicycle rentals and offers tours with local guides. Tourists can explore the village independently of the organised tours. A three-hour bike trail tour allows for village exploration, including carpentry workshops, ship building areas, ancient houses, temples and a pagoda along the Thu Bon River. The tour highlights are stops to visit artisanal workshops where visitors learn the family history of the artisans, their production techniques and procedures while observing skilful carpenters making souvenirs, and buying souvenirs direct from the producer (Hoi An Tourist Information Center, 2007). Revenue generated from entrance fees and tourism service was approximately 15,000 USD in 2006, as revealed in the informal talk with the Director of the Craft Centre.

The diversification of Kim Bong, from manufacturing wooden articles to souvenir supplier and later to tourism attraction, was initiated by public, private and non-governmental stakeholders (International Trade Centre, 2005). Initiatives aimed to increase employment and income through tourism and craft sales, managed by a tourism cooperative. Stakeholder roles are described below.

First, governance and tourism promotion is undertaken by the public sector, including central, provincial and district and community government. At a national level, the Vietnam Administrative of Tourism oversees coordination and promotion of the destination as a traditional handicraft village. This is in line with the national marketing strategy emphasis on handicraft villages as national cultural heritage (Viet Nam Administration of Tourism, 2004). At the provincial level, the Tourism Department of Quang Nam Promotion Centre promotes the development of handicraft village tourism around Hoi An. At a district level, Hoi An Commerce and Tourism Division distribute village tour information in Hoi An Tourist Centre. At the community level, Kim Bong and Hoi An Peoples Committee invested in tourism infrastructure, such as Craft Centre, the wharf and biking trails. The local community, led by Kim Bong Peoples Committee, has formed a cooperative to operate the souvenir production and tourism activity.

Second, the private sector has an important role marketing and developing products. Private businesses are the main channels bringing products to market. The interview with an official, who was in charge of tourism promotion for Hoi An Ancient Town and surrounding areas, revealed that private businesses such as tourism companies play active roles in promoting the destination. Numerous tourist companies include Kim Bong as an optional tour within package tours to Hoi An Town, generating visitation to the village. However, promotional activities are fragmented and subject to business demands rather than the benefit of the community. For example, package tours often limit the time to visit the village to about half an hour and do not include biking tours in the itinerary.

Third, the role of both national and international non-governmental organisations (NGO) is significant in promoting tourism-related products of the region. In particular, Quang Nam Small and Medium Enterprise and Cooperative (COOPSME) provide technical support, such as managerial training for the community cooperative leader. International NGOs, such as UNESCO and WWF, assist with field research and assistance for environment sustainability. An international NGO, the International Trade Centre, provides experts to assist the implementation of tourism site development while local government invested in Kim Bong infrastructure resulting in the Craft Centre construction (International Trade Centre, 2005).

Discussion

The cases of Bat Trang and Kim Bong demonstrate how household man-ufacturing businesses of traditional handicraft craft villages have success-fully engaged with the transition from a socialist to a market economy, creating entrepreneurial solutions in the supply of souvenirs. The two cases also illustrate characteristics of regional development of the peripheries. Tourism souvenirs have played an intermediary role within this transitional process, connecting traditional craft making to businesses in the tourism industry. Tourism souvenirs not only represent the symbolic meaning of places (Gordon, 1986), but the places of making these products have been shown to gradually become destination attractions (Moscardo, 2004). A number of factors discussed below contribute to this transition.

Location

Both villages are located close to tourist-generating centres (Hanoi and Hoi An). The short distance from these centres to the villages allows for day trips. The extension of tours to these peripheral areas meets a demand to lengthen tourist stay and diversify tourism products. In addition, alternative means of transportation are available. Bat Trang can be visited by public bus from the city centre, while a boat trip from Hoi An to Kim Bong offers vari-ety from the traditional tourist coach. Development of locations on the periphery allows craft outlets to spread out towards the tourist-generating centres, as Cohen (1995) found in his studies of village/urban craft ribbon developments in Thailand.

Handicraft traditions

Vietnam's handicraft villages have a reputation for skilled work and authentic handicrafts (Japan International Cooperation Agent, 2004). Both villages studied are known for traditional handicrafts, thus, having high aes-thetic amenity values (Hall & Muller, 2004), a favourable condition to develop tourism-related products. A characteristic of the traditional handi-craft product is that it has an artistic base and the potential to be a tourism souvenir (Wilkins, 2011). The role of a Ceramic Market (Bat Trang) and a Craft Centre (Kim Bong) where handicrafts are introduced to tourist visitors is essential. Informal education is important in motivating travellers to visit workshops and to purchase souvenirs during visits to the villages.

As Cohen (1995) found in his study of craft villages in Thailand, both of these villages in Vietnam face the same challenges of how to be competitive in producing and pricing an adequate variety of products. Traditional techniques of hand-made souvenirs are often more expensive and time-consuming than mass-production counterparts. In particular, threats come from Chinese imported products. The shortage of labour and possible loss of the transfer of intergenerational knowledge and skills is another challenge, as the younger generation lacks interest in continuing their family handicraft traditions.

Stakeholder involvement

A successful transition from production to a destination attraction requires the support of various stakeholders, as Hall and Jenkins (1998) stated about the interventionist role of the national and local states. In both cases, diversification has been influenced by the national policy to develop handicraft villages as tourist attractions (Viet Nam Administration of Tourism, 2004). In addition, private sector tourism businesses also play an active role. Needing to diversify tourism products, tourist companies look for new destinations in peripheral areas. In addition, NGO consultants were actively involved in the community-based tourism project in Kim Bong, playing a significant role in providing advice for product development. In Bat Trang, however, the NGO did not participate in development, since the village already had a relatively well-developed ceramic trading business and the villagers were not considered to need assistance for improving livelihoods. The role of local entrepreneurs and tourist companies, thus, has been the driver of the transition to tourism services in Bat Trang.

Transition to tourism

The evolution of the two villages from handicraft manufacturing to souvenir suppliers, and lately to tourist attractions, reflects the transition from a centrally planned to a market-oriented economy. Both villages illustrate the outcomes of a Vietnamese export-led development strategy (Japan International Cooperation Agent, 2004). Transitioning from production to tourism attraction, the villages reflect local innovation underpinned by government policy. However, each has taken a different route and emphasis. Bat Trang developed as a tourist destination much earlier than Kim Bong. The operation of tourism services in Bat Trang by individual local entrepreneurs was initiated in the late 1990s. In contrast, tourism services in Kim Bong had been introduced adjacent to the itinerary of Hoi An in the late 2000s, managed by a tourism cooperative. Yet both villages have managed, by different means, to maintain sustainable employment in urban and town peripheries.

Destination attraction

As living and working handicraft villages, Bat Trang and Kim Bong are emerging as destination attractions for tourist shopping experiences. This is evident in tours incorporating visits to these villages and in the presence of independent travellers. Being located on the peripheries of the urban destinations of Hanoi and Hoi An is significant, not only in attracting tourists, but in the ability of the villages to supply the more urban destinations with souvenirs. While Bat Trang attracts tourists by offering participative experience in ceramic production, Kim Bong emphasises its reputation in wood carving, displaying traditional production techniques in a peaceful rural setting.

Conclusion

This chapter has identified the gradual transition of Vietnam's handicraft villages from centres of production to destination attractions on the peripheries of major destinations. Factors of location, handicraft traditions and stakeholder involvement have been found to be significant in this evolution. The convergence of an increase in international and domestic tourism and a transitional economy has propelled an increased interest in the handicraft villages as sites for souvenir production and destination attractions for tourism. Recognising this evolution and a coming of age for these villages, is their promotion as part of the national cultural heritage of Vietnam (Viet Nam Administration of Tourism, 2004). Existing on the periphery in a developing country context, the villages producing local hand-made souvenirs are influenced by the globalisation of both tourism and trade, as tourism has brought increased visitation, but increased demand for souvenirs has led to pressures to use global suppliers. As the villages are still mainly craft production centres at an early stage of their involvement with tourism, continued research could focus on their evolution as destination attractions in a glocal context and their role in further developing handicraft village tourism in Vietnam.

References

Archer, B. (1998) Tourism and island economies: Impacts analysis. In C. Cooper (ed.) *Progress in Tourism, Recreation and Hospitality Management*, Vol. 1 (pp. 124–134). London: Belhaven Press.

Bryant, J. (1998) Communism, poverty, and demographic change in North Vietnam. *Population and Development Review* 24 (2), 235–269.

Chang T., Milne, S., Fallon, D. and Pohlmann, C. (1996). Urban heritage tourism: The global–local nexus. *Annals of Tourism Research* 23 (2), 284–305

CK&T Ceramics Company (2006) Bat Trang Village. Accessed 10 March 2012. http://sites.google.com/site/cktcrafts/battrangvillage

Cohen, E. (1995) Touristic craft ribbon development in Thailand. *Tourism Management* 16 (3), 225–235.

Dahle, H. (2010) Tourism, small enterpreises and community development. In D. Hall and G. Richards (eds) *Tourism and Sustainable Community Development* (pp. 154–168). London: Routledge.

Dang, T.L. (2008) Tourism development in Bat Trang ceramic village. Honors dissertation, Hanoi Tourism College, Hanoi, Vietnam.

Fairhurst, A., Costello, C. and Fogle Holmes, A. (2007) An examination of shopping behavior of visitors to Tennessee according to tourist typologies. *Journal of Vacation Marketing* 13 (4), 311–320.

Friends of Vietnam Heritage (2006) *Bat Trang Self-guided Walk*. Hanoi: The Gioi Publishers.

Friends of Vietnam Heritage (2010) *Hoi An Ancient Town*. Hanoi: The Gioi Publishers.

German Technical Cooperation (GTZ) (2007) *A consultant report on survey of craft village in Quang Nam*. Hanoi: Vietnam.

Getz, D. (1993) Tourist shopping villages: Development and planning strategies. *Tourism Management* 14 (1), 15–26.

Giang, N.T. (2009) *The Social Impacts of the Global Economic Crisis on Two Craft Villages in Viet Nam*. Vietnam: Oxfam GB.

Goodman, J.E. (2005) *Uniquely Vietnamese*. Hanoi: The Gioi Publishers.

Gordon, B. (1986) The souvenir: Messenger of the extraordinaory. *Journal of Popular Culture* 20, 135–146.

Hall, C.M. (2005) *Tourism: Rethinking the Social Science of Mobility*. Harlow: Prentice-Hall.

Hall, C.M. (2007) North–South perspectives on tourism, regional development and peripheral areas. In D. Muller and D. Jasson (eds) *Tourism in Peripheries: Perspectives from the Far North and South*. Wallingford: CABI .

Hall, C.M. and Jenkins, J. (1998) Rural tourism and recreation policy dimensions. In R. Butler, C.M. Hall and J. Jenkins (eds) *Tourism and Recreation in Rural Areas* (pp. 19–42). Chichester: John Wiley.

Hall, C.M. and Muller, D. (2004) *Tourism, Mobility and Second Homes: Between Elite Landscape and Common Ground*. Clevedon: Channel View Publications.

Hampton, M. (2005) Heritage, local communities and economic development. *Annals of Tourism Research* 32 (3), 735–759.

Healy, R.G. (1994) Tourist merchandise as a means of generating local benefits from ecotourism. *Journal of Sustainable Tourism* 2 (3), 137–151.

Hoi An Tourist Information Center (2007) *Kim Bong Carpentry Village Brochure*.

International Trade Centre (2005) *Poverty Reduction Through Community-Based Tourism in Vietnam: A case Study*. Hanoi, Vietnam: ITC Export-led Poverty Reduction Program Team (EPRP).

Japan International Cooperation Agent (JICA) (2004) *The Study on Artisan Craft Development Plan for Rural Industrialization in The Socialist Republic of Viet Nam*. Hanoi, Vietnam: Japan International Cooperation Agency.

Littrell, M.A. (1990) Symbolic significance of textile crafts for tourists. *Annals of Tourism Research* 17 (2), 228–245.

Littrell, M.A., Anderson, L.F. and Brown, P.J. (1993) What makes a craft souvenir authentic? *Annals of Tourism Research* 20 (1), 197–215.

Littrell, M.A., Baizerman, S., Kean, R., Gahring, S., Niemeyer, S., Reilly, R., *et al.* (1994) Souvenirs and tourism styles. *Journal of Travel Research* 33 (1), 3–11.

Luong, P.T. (2006) Support for local and cottage industries. In *VNAT/APEC/SME Seminar*, 21 September 2006, Hanoi, Vietnam.

Ministry of Argriculture and Rural Development (2002) *Decree on recognition of traditional handicraft, and handicraft villages.*

Moscardo, G. (2004) Shopping as a destination attraction: An empirical examination of the role of shopping in tourists' destination choice and experience. *Journal of Vacation Marketing* 10 (4), 294–307.

Nguyen, Q.N. (2009) From craftvillage to clusters in Vietnam: Trasition through globalisation. In *Asian Industrial Clusters, Global Competitiveness and New Policy Initiatives.* Singapore: World Scientific Publishing.

Rutten, R. (1990) *Artisans and Entrepreneurs in the Rural Philippines: Making a Living and Gaining Wealth in Two Commercialised Crafts.* Amsterdam: VU University Press.

SNV (2007) *Community-Based Tourism: Lesson Learnt from Viet Nam.* Hanoi, Vietnam.

Sarkun, T. and Banerjee, S. (2007) Artisan clusters – some policy suggestions. *Innovation Journal: The Public Sector Innovation Journal* 12 (2), 1–14.

Swanson, K.K. and Timothy, D.J. (2012) Souvenirs: Icons of meaning, commercialization and commoditization. *Tourism Management* 33 (3), 489–499.

Szydlowski, R.A. (2008) Expansion of the Vietnamese handicraft industry: From local to global. Masters of Arts, Ohio University.

Timothy, D.J. (2005) *Shopping Tourism, Retailing and Leisure.* Clevedon: Channel View Publications.

Teo, P. and Lim, L. (2003) Global and local interactions in tourism. *Annals of Tourism Research* 30 (2), 287–306.

Turner, L. and Ash, J. (1975) *The Golden Hordes: International Tourism and Pleasure Periphery.* London: Constable.

Tran, N.C. (2003) *Presentation for e-Awareness Seminar Series for Asian Parliamentarians.* Hanoi, Vietnam.

Vasavakul, T. (2001) Vietnam: Doimoi difficulties. In J. Funston (ed.) *Government and Politics in Southeast Asia* (pp. 372–409). Singapore: Institute of Southeast Asian Studies.

Viet Nam Administration of Tourism. (2005) *Tourism Marketing Plan – Vietnam 2008–2020.* Vietnam National Tourism Administration.

Viet Nam Administration of Tourism. (2010) *Tourism Statistics 2009*, accessed 31 March 2012. http://www.vietnamtourism.gov.vn/english/index.php?cat = 0120

Viet Nam Administration of Tourism. (2012) *Tourist Statistics 2011*, accessed 31 March 2012. http://www.vietnamtourism.gov.vn/english/index.php?cat = 0120

Weaver, D. (1998) Peripheries of periphery: Tourism in Tobago and Barbuda. *Annals of Tourism Research* 25 (2), 292–313.

Wilkins, H. (2011) Souvenirs: What and why we buy. *Journal of Travel Research* 51 (2), 239–247.

Yin, R.K. (2009) *Case Study Research.* London: Sage.

12 World Heritage-themed Souvenirs for Asian Tourists in Macau

Hilary du Cros

Souvenir production and collection are commonplace aspects of tourism activity. Cohen (1993: 6) has stated 'the commodification of the past through the antiquities markets for tourists and collectors has helped to shape the growing heritage industry.' That is, the cultural heritage of most places became accessible to tourists by the later 20th century in that it became fodder for souvenirs. However, Hitchcock (2000: 1) observed at the beginning of an edited book on the material culture of tourism 'items purchased on a holiday are meaningful and are often more than simple mementos of time and place'. If this is true for all tourists, what do Asian tourists consider as mementos and what do they consider meaningful? What would they take home to keep or give away as gifts to signify that they have visited a place?

Previous studies of Asian heritage-themed souvenirs have often concentrated on aspects of commodification and authenticity in relation to their appeal to non-Asian tourists. These tourists would consider arts and crafts they buy as souvenirs to be the cultural products of 'other' and somehow sacred in that they are perceived to be an authentic memento of a vanishing world (MaCannell in Hitchcock, 2000). This would be the case still, if one was considering the behaviour of a small number of purposeful cultural tourists (McKercher & du Cros, 2002) who are seeking a souvenir of a deeper experience of local culture, rather than sightseeing tourists bent on fast easy consumption of a destination's culture. Ideally, it is most advantageous to a destination to cater to both markets in some way.

Many articles on Asian tourism only mention souvenirs briefly in passing when considering the impacts of mass tourism on local communities (for instance Arlt & Xu, 2009; Leung, 2008). There is a new approach that blends sociological observation with postmodernist viewpoints. Over the last few years, Winter (2009a) has been teasing out strands of enquiry in a way that is having an impact on cultural tourism studies of Asian tourists. His most significant work deals with Angkor World Heritage Site (WHS) in Cambodia (Winter, 2007, 2009b) and an edited book on Asian tourists (Winter *et al.*, 2009). Fundamentally, he believes that Asian modernity is changing the nature of cultural tourism in the region away from that developed primarily to service Western cultural and sightseeing tourists, towards that more strongly favouring the needs of the regional market.

Twenty-first century academic studies are beginning to identify how the impact of domestic and regional tourism associated with developing countries, such as China, Thailand and India among others, is bringing new insights on tourism (refs from work). Of particular interest is how increased attention to Asian tourism will inevitably bring a new examination of the demand and supply of heritage-themed souvenirs in relation to WHSs in Asia as there is mounting interest in these heritage assets from the viewpoint of this emerging market in the context of local responses to global change. This chapter grew out of an interest in the developmental history of the key heritage tourism precinct in Macau, St Paul's Ruins, as a cultural district incorporating arts and leisure activities for a new class of Asian cultural tourists. The Ruins are a visually dramatic aspect of this WHS and an archaeological site that is an iconic representation of the historic religious influence of Macau in the region going back to the mid-1500s. The precinct has been chosen for this study by the author because of its popularity. During the year after inscription, nearly one in 14 tourists visited this part of the WHS (du Cros, 2009).

The Ruins also appear in the tourist logo for Macau and in many examples of tourist promotional media. It is therefore a must-see attraction with high visibility and was considered a heritage icon before being inscribed as an element of the WHS. The current study of souvenirs was conducted within the pedestrianised area of the district surrounding the Ruins over three years in two stages; from March 2008 to June 2009, and a latter one from June to December 2011. The study gathered information from over 36 brief semi-structured interviews with shop-managers, tourists, craftspeople, artists and local designers of tourism products. Observations were made every two months over these two periods in a series of visits to the district that mapped the location of popular certain shops and their location in relation to others, the new shops opening and the old ones closing down, reducing space or changing merchandise. This mixed approach of interviews and observations was chosen to identify developmental and spatial indicators of how the Ruin's souvenir shops were changing in response to a possible market shift. It also explored whether a perceived market shift is reflected in the material culture of tourism – particularly the more recently developed interest in food souvenirs – evident in Macau Special Administrative Region (SAR) since the inscription of the Historic Centre of Macau on the World Heritage List (WHL) in 2005. The packaging often displays images of the Ruin's and other heritage elements of the Historic Centre, and the shops contain products based on a cuisine that is representative of Macau's fusion culture that is celebrated in the statement of Outstanding Universal Values (OUVs) for which the Site was inscribed.

Importance of Theming World Heritage Site Souvenirs

Each year the United Nations Educational, Scientific and Cultural Organization (UNESCO) World Heritage Committee meets to discuss the inscription of new natural and cultural heritage sites of OUV on the WHL. 'The Historic Centre of Macau' was declared a WHS in July 2005, amidst much fanfare, most particularly by the Macau Government Tourism Office (MGTO). Authorities such as, Shackley (2006), Hall and Piggin (2003), Boyd and Timothy (2006) and Leask and Fyall (2006), have indicated that WHSs are more likely than any other type of heritage asset to become primary attractions and that marketing and management regimes, implemented after inscription, should reflect the associated increase in visitation accordingly.

Hall and Williams (2008: 195) note that brand innovation 'can refer to new ways of reinforcing existing brands, the development of new ones and new means to protect the values and designs of existing brands'. Macau is an interesting case to examine in relation to this issue as it has welcomed modern tourism since the 19th century with day-trippers arriving from Hong Kong and has long been a hub for East–West cultural hybridisation. Theming souvenirs in this context would mean that the design and marketing of souvenirs would use elements of Asian and non-Asian techniques and experience in order to appeal to regional tourist markets as much as those further afield. However, there has been no discussion in academic literature regarding the importance of theming and World Heritage designation to the development of local souvenir industries. This area appears to be completely unexplored with the exception of the vague mention of souvenir production using materials from natural World Heritage areas in Tasmania by Hume (2005).

Macau as an Emerging Regional Centre for Tourism

Macau has undergone its most intense tourism development in the last ten years with arrivals jumping from 9 million to over 24 million (DSEC,

2012; MGTO, 2006, 2012) It is located one hour's distance by ferry from Hong Kong and immediately adjacent to the southern Chinese province of Guangdong. This gives Macau a unique position in relation to Asia, both now and in the past (du Cros & Lee, 2007). Macau was returned to Chinese control from Portuguese in December 1999. It is governed as a SAR within China. Macau's development was greatly assisted in mid-2003 by the Central Government's policy to relax visa requirements for leisure travel from much of southern China to it. By allowing independent as well as tour-group visas to be issued to residents in certain cities and towns in China through the Independent Visit Scheme (IVS), Macau's tourism flourished quickly. Since then, the Macau Government's vision is for Macau to become a key international destination for recreation, Meetings Incentives Conferencing and Exhibitions (MICE) and shopping destination, with less reliance on gaming as time goes by. Cultural and sport tourism are recognised special interest tourism niches for Macau (du Cros, 2009; Ung & Vong, 2010).

Chinese Mainland tourists for April 2012 comprised more than 61.6% of the total visitors coming to Macau (DSEC, 2012). Since 2010, the number has hovered around 60% with many of these visitors being cross-border arrivals from neighbouring Guangdong. The China outbound travel market has been increasing for both Hong Kong and Macau SARs since 2003, with only some recent drops owing to the international financial crises and a temporary change in 2008 to the IVS as a result, possibly to keep expenditure local though this was not expressly stated. It should be recognised that the demand has the potential to increase for more special interest tourism products as it needs diversify with greater affluence and education, as is already occurring with tourists from the larger cities (Xiang *et al.*, 2009).

Overall, it has been observed that it is likely a continuum exists in Macau that has gambling addicts at one end (people will not leave the casino at all during their stay) and purposeful cultural tourists at the other end (who will look at every exhibit in a museum). The worlds of these two extremes do not meet at all and they seem to be able to co-exist for now at the destination contentedly. Significantly, there is likely to be a middle group, whose motivations regarding cultural tourism are more variable, and which could take into account a greater range of factors beyond time and interest (du Cros, 2009). Knowing more about how this group can be encouraged to engage more with the culture of Macau is going to be crucial to the survival of its cultural heritage in the future. It will also be important for understanding how global, regional and local influences affect cultural tourism development in Asia, particularly regarding demand, supply issues and best practices that could embrace souvenirs. Local Macau food producers and retailers response to increased demand for souvenirs was observed over the course of the study.

Results

Accordingly, field visits for this study gauged the popularity of certain shops and found that food versus non-food souvenirs dichotomy has meant that the former have greatly increased in number and the latter are losing ground. Interviews with shopkeepers confirmed that this change has only taken place in the last five years. It is also evident in some tourist blogs that perishable gifts, such as the products of the Macau bakeries, are considered suitable souvenirs of Macau (Joyful Steps, 2011). This trend has allowed some of the most popular shops to set up franchises and spread their products throughout other tourist precincts in Macau and in hotel casino retail areas. Some shops are also bakeries with the food cooked in front of customers and where free samples are aggressively offered. Other shops are merely outlets, such as those in the large hotel casino megaresorts. Much of their success is owing to many of the local bakery franchises trying more international marketing practices than before to promote their products to regional Asian tourists, such as the use of celebrity endorsements and elaborate packaging. Key bakery chains have adopted packaging with a similar approach to aesthetics developed originally in Japan. For an example of how Macau is using Western marketing principles, the Japanese approach to aesthetics and its heritage can be seen in the promotion of the bakery products as food souvenirs, at the websites of the two key bakery franchises, Koi Kei Bakery (2011) and Choi Heong Yuen Bakery (2011).

The top selling souvenir products in the study were 'wife' pastries, almond 'coin' biscuits and egg rolls with the name 'St Paul's Ruins' on the packet, however, postage stamps and casino chips with the Ruins on them are also very popular, particularly with the Mainland Chinese tourist market (see Table 12.1). The excessive use of the St Paul's Ruins image seems to be employed to distinguish Macau-bought souvenirs from those bought on trips elsewhere for tourists. This is similar to the usage of other famous WHS icon images overseas (e.g. Eiffel Tower on t-shirts). It also links the tourism icon with the WHS brand in the consumers' minds. The mass-market sightseeing tourists will no doubt continue to buy these products for some time to come.

The shop owners explained in the interviews that there had been a strong market shift attendant on the rise in Mainland Chinese tourist numbers to Macau. Some owners acquainted the drop in popularity of Chinese non-food souvenirs (furniture, crafts and other accessories) with increased ease of access to China for expat and Western international tourists who can now buy these items directly from source in Zhuhai or Shenzhen instead of relying on Macau homewares and crafts shops as middlemen. The increase in tourists with Chinese cultural backgrounds had seen food souvenirs

Table 12.1 Souvenirs ranked in order of popularity

Souvenir type/use of WHS images	Market	Popularity	Authenticity	Relevance to Macau's cultural identity
Chinese cakes, rolls and biscuits (heavy use of St Paul's Ruins image)	Tourists from Mainland China, Taiwan, Singapore and Hong Kong	Highly popular gift souvenir on return. Requested by friends/relatives as gift. Shopping bags (WHS image) kept and reused	High	Low: most recipes from Western side of Pearl River from 1930s. Intermixed with Portuguese
Portuguese tarts (not replicas). Heavy use of St Paul's Ruins image	All tourists	High: as above	High	Moderate: Macanese with Portuguese and English influences
Small cheap mass-produced items, e.g. fridge magnets, postcards, key rings, casino chips (St Pauls Ruins)	All tourists, but mostly Chinese tourists	High: reasons given include cheap, cute, easy to transport,	Low: all made across the border	Images and sometimes designs
Stamps, coins, first day covers, etc.	Chinese and other Asian tourists	Moderate to high: stamps are particularly popular as gifts to children	Moderate to high: most are produced in Macau for special events, etc.	Moderate to high: most have been used or made for Macau
Portuguese style small mass produced – rooster statues, football posters, flags, shoes, posters, flags on key rings, catholic religious icons, etc.	All tourists, but mostly mainland Chinese tourists	Moderate: reasons given include cheap, cute, easy to transport	Moderate: either made across the border or back in Portugal	Low: images and designs are the same as souvenirs found in Portugal

(Continued)

Table 12.1 (Continued)

Souvenir type/use of WHS images	Market	Popularity	Authenticity	Relevance to Macau's cultural identity
Small Chinese antiques/handicrafts	Western expat, international tourists. South East Asia, Chinese diaspora	Moderate: small pieces are easy to transport	Low to moderate	None: almost all are made across the border
Chinese kites and other toys	All tourists	Low to moderate: mostly popular for price and design	Low to moderate: kites are not mass produced	Low: only kites are sometimes made locally
Casual fashion clothes and accessories (produced and designed locally)	Young Chinese and Asian tourists	Low to moderate: only a recent initiative by two young designers	Moderate: much merchandise made locally	Moderate: some effort to refer to local history and culture stylistically
Paintings, sculptures, posters and other artworks	Expat, Western tourists (originals). All tourists (copies)	Low to moderate: many are poorly presented and marketed or the galleries are not easy to locate	Moderate: most locally produced. Copies of popular Western paintings painted to order	Low to high: local artists working with local themes. More common
Chinese antique/reproduction furniture	Western expat and international tourists	Low: too familiar for Chinese tourists, low interest in antiques, costly to transport home	Low: Chinese concerns about authenticity –reproductions	None: can also be found across the border
Chinese traditional clothes (e.g. cheongsams)	Western expat and international tourists	Low: not many stay long enough for fittings, etc. these days	Moderate: the older tailors still make traditional way	Low: similar designs in Hong Kong and Guangdong

grow in popularity instead (see Table 12.1). It is likely that perishable gifts rather than long-term keepsakes are more often purchased by Asian cultural tourists, particularly those originating from Mainland China and Hong Kong, the two key markets for Macau, as they are easy to carry and pleasing to recipients.

That such a wide range of tourist shops exists close to St Paul's Ruins shows it is considered an important hub by the local populace, tourists and destination marketers. The results of the interviews suggest that it has always been important to tourism even before the World Heritage inscription and a pattern has emerged, see Table 12.2.

Also, it was observed that there are many repeat visitors to the food souvenir shops and these tourists are not looking for long-lasting mementos to take home. Nor are these purposeful tourists and who might be searching for educational and authentic souvenirs of Macau. During the study, the search for authentic Macau souvenirs of arts and crafts yielded only a few

Table 12.2 Changing pattern of supply and demand

Period	Date Range	Comments
Colonial Portuguese	Before 1980s	Hong Kong, British and Portuguese tourists buying mainly locally made woodwork, furniture and small cheap souvenirs
Transition/early postcolonial	1980s to 2003	Same market but with the addition of more Japanese and Taiwanese tourists and Mainland Chinese tour groups showing an interest in the Chinese food souvenirs. Shops selling these were not close to the Ruins, however
Pre-crisis postcolonial	2003 to 2008	Impact of Independent Visa Scheme means more repeat visits and more cross-border tourism from China. The dependence on the Chinese mainland market by souvenir sellers has meant that food souvenirs bakery shops have multiplied and migrated closer to the Ruins. Rents have likewise risen making it difficult for older businesses that do not own their premises to compete
Current postcolonial	2008 +	More businesses that cater to tour groups rather than independent tourists as these are not always a reliable option. More Portuguese-themed souvenirs available

items (see Table 12.1) compared with what the author remembers from before 2004. For instance, there are just two companies still running that comprise the industry of wooden religious carving in 2009: 'Tai Cheong Wood Engraving Co.' and 'Artigos Religiosos Kuong Weng' (Kuong Weng Religious Figure Carving and Woodwork). Upheavals in Southern China have not changed the traditional values associated with the care in detail and authenticity of these articles in Macau according to tourists, practitioners and craftspeople interviewed. Macau has a greater continuity of cultural traditions in the last 100 years than China, as it has not suffered through the Cultural Revolution and that has enhanced positive feelings about the intangible heritage attached to these items. However in mid-2011, one of the companies was forced to close its shop close to the Ruins owing to rising rents. Since then, it has been more difficult for tourists to access their products.

The other non-mass produced category of souvenirs studied was that of art. Closest to St Paul's Ruins is St Paul's Corner Gallery. It opened late in 2007 in a protected historic Portuguese-style building. Local artist and one of the organisers, James Chu, was concerned that the gallery should tap into an art market in Macau and overseas. However, Chu has commented, 'One of the heaviest burdens that Macau artists have to carry is their own label' (Chu in Kok, 2008: 113). He feels both the artists and the markets would benefit from more exposure of Macau art outside of its familiar setting in Macau. Chu's hope is that the gallery's efforts to do this at art fairs in the region will draw art tourists to Macau to buy from the centre of production at lower prices, while having an enjoyable leisure experience (Kok, 2008).

On the other side of St Paul's Ruins from the main shopping area, the St Lazarus Conservation Area of Portuguese-style buildings has finally started to attract galleries, boutiques, restaurants and cafes in small numbers, despite a slow start. The Sao Lazaro Creative Industries District Plan was completed in 2000 for the area and has, to some extent, been integrated into local planning for its protection. Close to Christmas in 2011, it also ran a programme of street performances featuring Portuguese-style circus activities to attract tourists.

The entrepreneurial side of incorporating heritage-related motifs in souvenirs has been discussed only occasionally in tourism literature. The Banyan Tree Resort Group saw an opportunity to develop a souvenir line based on the Asian-themed handicrafts and toiletries that used local heritage motifs after many started disappearing from its rooms (Pritchard & Morgan, 2006). UNESCO and United Nations World Tourism Organization (UNWTO) have provided assistance in building capacity for community-based souvenir development, which can include conferences and publications on marketing and design, to increase competitiveness and tourist appeal UNWTO, 2008.

Also, there has been anecdotal evidence of exchanges between traditional handicrafts producers and individuals with international design experience that has aided both (Cave, 2009). Meanwhile in Macau, some new designer products with heritage themes show more promise for appealing to young Asian tourists, certainly more than the products available in the old furniture and crafts souvenir shops aimed at the Western expat and international tourist market. One example is the shop of a Portuguese designer, who moved here in 2007 from Lisbon and who uses local architectural motifs on her printed fabrics. One of her fabrics uses the patterns made by the ventilation holes on folding metal doors of the older local shops and has been made into dresses and furnishings. She is looking at Macau with fresh eyes, much in the way that a purposeful tourist might, and finding inspiration for her designs that are likely to appeal to young urban tourists from the region. Another example is a young local Chinese designer, who briefly established four shops in Macau selling her own products to tourists. She noted in an interview that even though she did not have any formal marketing plan, she aimed her merchandise at 'trendy young tourists', who are much like herself. The range she personally designed under the brand name of Ao2 included shopping bags, baseball caps, handbags, t-shirts and purses. The cloth shopping bags were notable in being influenced by Macau's maritime history with fishlike designs and boats. The colour of the fabrics, the type of products and the shops' presentation show recent Asian and international marketing and design influences. She ran the shops until most of her merchandise was sold and then use the proceeds to establish herself overseas. Unfortunately, this was a common pattern for young designers unwilling or unable to find an on-going foothold in Macau.

Finally, the government has taken an interest in heritage-themed arts and crafts souvenirs. In November 2011, the MGTO dropped the rents in a restored historic Portuguese building at the foot of the steps to the Ruins, known colloquially as the 'Yellow House'. Portuguese and local Macanese designers and artists were encouraged to sell their wares on different levels in the building, including next to a Portuguese café. This initiative may yield some positive results for the creative industries of Macau, particularly as more cultural exchange will occur as artists migrate to Macau from Portugal, given the economic problems in Europe and Macau's postcolonial ties.

Discussion and Conclusion

The integration of international marketing practices and the creation of recognised local brands of designer souvenir franchises are not new in other cities in Asia, as an interesting tension develops between what is recognisable as global and what is uniquely local plays out in the goods on offer. The Jim Thompson (Thai silk company) brand proved that such goods could command admiration. Also, Hong Kong's local Chinese designer brands, Shanghai Tang and G.O.D. (Goods of Desire), have taken it to another level, trying to promote an Asian design sensibility that appeals to the modern Asian consumer and others (du Cros, 2010). What has not been tried anywhere is a line of merchandise that ties this approach to WHSs for either the young or purposeful Asian souvenir purchaser with a critical eye. The type of commodification required of these souvenirs will be different to that which attracted the non-Asian tourists discussed in earlier works on souvenirs by Cohen (1993) and Hitchcock (2000).

What does all this say about Asian modernity as posited by Winter (2007, 2009a) and World Heritage-themed souvenirs? The situation is in a state of flux as most young Asian tourists to Macau appear to be collecting experiences without any lasting physical reminder to mark their trips, such as indicated by Joyful Steps (2011), because they are mainly buying food souvenirs. For frequent visitors one might expect this, however, there appears to be an opportunity arising to highlight to less frequent tourists the arts near the Ruins and within the St Lazarus Conservation Area. This approach could also be extended to revitalise the local woodcarving, kite making and other crafts as well. Unfortunately for these local crafts, the current focus may be on Portuguese aesthetics as Asian tourists, without the opportunity to visit Portugal, currently take back a taste of it from Macau with the souvenirs depicting the WHS elements with Portuguese influences, items available from the newly arrived Portuguese artists and items sourced directly from Portugal.

Winter (2008) is correct in his view that Asian modernity is changing cultural tourism in the region away from that developed primarily to service Western cultural and sightseeing tourists as this market declines in numbers. Rising incomes in much of Asia and economic uncertainty at the point of origin for many Western tourists has increased the regionalisation of Asian tourism. Although popularity levels of souvenirs and the use of WHS architectural motifs in their marketing may be particularly common in Macau, it is likely that modernity in Asia has a more chameleon-like nature than that in the West. Accordingly, attractive souvenirs for many of these new Asian tourists to Macau need to be themed or designed in such a way as to also address their aspirations regarding future travel, fashion and cultural identity. WHSs are usually a primary tourist attraction and there is strong need to understand how local producers and retailers can support *local* talented designers, artists and marketers, as well as the

Portuguese ones, in attracting consumers more broadly from the regional market. It would also mean that more souvenirs could be produced that evoke a more meaningful memory of a place's unique World Heritage cultural values for these tourists, and one that will last longer than that from perishable food souvenirs, despite carrying the St Paul's Ruins image on the packaging.

References

Arlt, W.G. and Xu, H. (2009) Tourism development and cultural interpretation in Ganzi, China. In C. Ryan and H. Gu (eds) *Tourism in China. Destination, Culture and Communities* (pp. 168–181). London: Routledge.

Boyd, S.W. and Timothy, D.J. (2006) Marketing issues and World Heritage Sites. In A. Leask and A. Fyall (eds) *Managing World Heritage Sites*. (pp. 55–68). London: Butterworth-Heinemann.

Cave, J. (2009) Between World Views. Nascent Pacific Tourism Enterprise in New Zealand. Unpublished PhD thesis, Unversity of Waikato, New Zealand, accessed 27 April 2012. http://researchcommons.waikato.ac.nz/bitstream/handle/10289/3281/thesis.pdf?sequence=1

Choi Heong Yuen Bakery (2011) Accessed 28 December 2011. http://www.choi-heong-yuen.com/current/index.php

Cohen, E. (1993) Introduction: Investigating tourist arts. *Annals of Tourism Research* 20 (1), 9–31.

DSEC (2012) Macau Statistics and Census Service, accessed 28 April 2012 for latest arrival figures. http://www.dsec.gov.mo/e_index.html

du Cros, H. (2009) Emerging issues for cultural tourism in Macau. In Chung, T. and Tieben, H. (eds) Macau ten years after the handover (Special Issue). *Journal of Current Chinese Affairs (China Aktuell)* 1, 73–99.

du Cros, H. (2010) Goods of desire: Visual and other aspects of western exoticism in postcolonial Hong Kong. In P. Burns, C. Palmer and J.A. Lester (eds) *Tourism and Visual Culture*, Vol. 1 (pp. 170–80). London: CABI International.

du Cros, H. and Lee, Y.S.F. (2007) *Cultural Heritage Management in China: Preserving the Pearl River Delta Cities*. London, Routledge.

Hall, C.M. and Piggin, R. (2003) World Heritage Sites: managing the brand. In A. Fyall, B. Garrod and A. Leask (eds) *Managing Visitor Attractions. New Directions*. (pp. 203–19). Oxford: Butterwork Heinemann.

Hall, C.M. and Williams, A.M. (2008) *Tourism and Innovation*. London: Routledge.

Hitchcock, M. (2000) Introduction. In M. Hitchcock and K. Teague (eds) *Souvenirs: The Material Culture of Tourism* (pp. 1–17). Aldershot: Ashgate.

Hume, D.L. (2005) An exhibition of art, craft and souvenirs from World Heritage Sites in Tasmania and Far North Queensland. In D. Harrison and M. Hitchcock (eds) *The Politics of World Heritage Negotiating Tourism and Conservation* (pp. 169–175). Clevedon: Channel View Publications.

Joyful Steps (2011) *Macau: Souvenir Shopping*, blog, accessed 12 December 2011. http://soo-sean.blogspot.com/2008/11/macau-souvenir-shopping.html

King, V.T. and Parnwell, M.J.G. 2011 World Heritage Sites and domestic tourism in Thailand: Social change and management implications. *South East Asia Research* 19(3), 381–420

Koi Kei Bakery (2011) *Introduction and Videos*. Accessed 14 December 2011. http://www.koikei.com/index.php

Kok, A. (2008) Tête-A-Tête with Beijing. Macau Artists' AFA Beijing Contemporary Art Centre Opens. *Macau Closer* November 2008, 108–13.

Leask, A. and Fyall, A. (2006) World Heritage Site designation. In A. Leask and A. Fyall (eds) *Managing World Heritage Sites* (pp. 5–19). London: Butterworth-Heinemann.

Leung, P.K.H. (2008) "The Southern Sound" Nanyin: Tourism for the preservation of traditional arts. In B. Prideaux, D.J. Timothy and K. Chon (eds) *Cultural and Heritage Tourism in Asia and the Pacific* (pp. 49–58). London: Routledge.

Macau Government Tourism Office (2006) Macau celebrates record-breaking 20 million arrivals, accessed 27 April 2012. http://industry.macautourism.gov.mo/en/pressroom/index.php?page_id=172&year=2006&sp=0&id=1676

McKercher, B. and du Cros, H. (2002) *Cultural Tourism: The Partnership between Tourism and Cultural Heritage Management*. Binghamton, New York: The Haworth Press.

MGTO 2012 Macau Tourism Industry Net. Visitor Arrivals. Accessed 5 March 2013: http://industry.macautourism.gov.mo/en/Statistics_and_Studies/list_statistics.php?id=39,29&page_id=10

Pritchard, A. and Morgan, A. (2006) Hotel Babylon? Exploring hotels as liminal sites of transition and transgression. *Tourism Management* 27 (5), 762–772.

Shackley, M. (2006) Visitor management at World Heritage Sites. In A. Leask and A. Fyall (eds) *Managing World Heritage Sites* (pp. 83–94). Oxford: Elsevier.

Sharpley, R. and Sundaram, P. (2005) Tourism: A sacred journey? The case of ashram tourism, India. *International Journal of Tourism Research* 7 (3), 161–171

Su, Ming Ming and Wall, Geoffrey (2011) Chinese Research on World Heritage Tourism Asia Pacific. *Journal of Tourism Research*, 16 (1), 75–88.

United Nations World Tourism Organization (2008), *Tourism and Handicrafts – A Report on the International Conference on Tourism and Handicrafts*. Madrid UNWTO.

Ung, A. and Vong, T.N. (2010) Tourist experience of heritage tourism in Macau SAR, China. *Journal of Heritage Tourism* 5 (2), 157–168.

Winter, T. (2009a) The modernities of heritage and tourism: Interpretations of an Asian future. *Journal of Heritage Tourism* 4 (2), 105–115.

Winter, T. (2009b) Asian destination: Rethinking material culture. In P.T. Winter and T.C. Chang (eds) *Asia on Tour: Exploring the Rise of Asian Tourism*. (pp. 52–66). London: Routledge.

Winter, T. (2007) *Post-conflict Heritage, Postcolonial Tourism: Culture, Politics and Development at Angkor*. London: Routledge.

Winter, P.T. and Chang T.C. (eds) (2009) *Asia on Tour: Exploring the Rise of Asian Tourism*. London: Routledge.

Xiang, R.L., Harrill, R., Uysal, M., Burnett, T. and Zhan X. (2009) Estimating the size of the Chinese outbound travel market: A demand-side approach. *Tourism Management* 31 (2), 250–259.

13 Lessons in Tourism and Souvenirs from the Margins: Glocal Perspectives

Lee Jolliffe, Jenny Cave and Tom Baum

Souvenirs are integral to global tourism, contributing to the experience of tourists by satisfying their need to take home a memento (Timothy, 2011). Shopping to acquire a local souvenir (consumption) is thus a major tourism activity. The souvenir may also be an item that is not purchased as a souvenir, but is kept as evidence of experience and as an aid to recalling the memory of the experience (Wilkins, 2011). Reflecting cultural heritage of local places, souvenir creation and production (supply) can contribute to enhancing livelihoods and economies (Healy, 1994), particularly for populations in peripheral destinations, or groups that are marginalised from other forms of economic activity (Rutten, 1990). This volume has reflected upon these aspects in terms of the glocal (combining the global and the local) construction of souvenirs of place, people and experiences.

With much previous research on tourism and souvenirs focusing on the material culture of what was bought (Gordon, 1986; Littrell *et al.*, 1993) and the motivations of why it was bought (Anderson & Littrell, 1995; Kim & Littrell, 2001), there was a need to move the discourse beyond the focus on the transaction of souvenir purchase in terms of choice and purchase intent. In this volume we have taken a broader perspective of the meanings of the souvenir, in particular with a glocal perspective at a range of peripheral local geographic locations influenced by global forces of tourism (for instance urban periphery, rural, village and small island destinations) and for groups at the margins of society who play a role in this tourism (such as migrant, ethnic, minority and indigenous populations).

This more expansive perspective of the investigation of the tourism souvenir phenomenon links into literature on the synergies between the local and the global, examined as the global–local nexus, often in an urban context (Chang *et al.*, 1996) and in alternative economies in worlds beyond the mainstream (Gibson-Graham, 2006) reflecting local lives that are influenced by globalisation (Katz, 2004). As Kaplan (1996: 145) notes 'against the homogenisation of globalisation, the qualities of the local or regional are frequently championed' arguing that while the global and local are spatial articulations of modernity, mapping these seemingly disparate spaces as a global–local nexus (glocal) cuts across many areas of study and cultural practices. Modes of production of globalisation include the reciprocal processes of globalised localisms, the process whereby particular locations are globalised (De Sousa Santos, 2004).

This concluding chapter reflects the central research themes of *glocalisation, impacts of glocalisation,* and *at the margins: people and places* identified in the introductory chapter, while referring back to the foundations of research on tourism souvenirs and glocalisation theories. In addition, the chapter considers the volume contents in the context of *tourism and cultural change,* the theme of the Channel View book series in which it is published. Research gaps and evidence are reviewed under the categories of *experience and behaviour, place and identity* and *case studies.* In the *discussion* we put forward a revised approach to theorisation in the field of tourism and souvenirs, incorporating aspects of glocalisation, its impacts and margins/peripheries, as well as theories of souvenir experience and behaviour, place and identity with the study of souvenirs. From this new perspective of souvenir studies, in the conclusion we extrapolate how this volume has moved ahead the *research agenda* for the topic of glocal expressions of souvenirs and sustainable tourism, noting areas for future research.

Research Themes

A number of research themes related to tourism souvenirs are examined within the volume. Themes include the theoretics of the glocal concept as a framework for analysis, an understanding of the impacts of glocalisation, theories of people at the margins and peripherality and practical perspectives of tourism and cultural change in relation to souvenirs.

Glocal

Using the theme of 'glocalisation' (representing the global–local nexus) allows for the analysis of souvenir contexts from the perspective of the local level juxtaposed with the influences of the global tourism industry. The process of souvenir creation (production) and purchase (consumption) is viewed essentially as a glocal phenomenon, combining the complexities of the global and the local. Demand for the souvenir as evidence of experience is global, driven by the globalised tourism industry reflecting the demand for global culture, yet the crafting of souvenirs as well as their initial acquisition and consumption at the point of purchase is local part of the complex and complicated global–local nexus whereby economic, cultural and environmental factors come together to nurture local development outcomes (Milne & Ateljevic, 2001). This complex interplay of forces, economic and social at both the global and local levels, places souvenirs clearly into a globalised yet glocal context.

Impacts of glocalisation

In the 21st century, the impacts of glocalisation are complex, since they are stimulated by transnationalism, internet connective technologies, pluralism, multiple modernities and different capitalisms driven by the global south. Thus evolving to the point that in time simultaneously glocal hybrid forms may emerge that affirm territorial difference as well as similarity (Pieterse, 1994). Diverse capitalist and non-capitalist economies of 'off the books' (Williams & Nadin, 2010) unwaged community-focused alternatives (Gibson-Graham, 2008, 2010) as lifestyle choices, transitional forms of enterprise and enterprise incubators, among others, may emerge.

The impacts of glocalisation processes are made manifest in the contexts, features and processes of everyday lives, including souvenir production, as glocal agents resist, respond to and interpret global influences (Katz, 2004). At local levels, glocalisation processes preserve and sustain craft traditions, cultural structures, community relationships and economies, as illustrated by the global contributions in this volume, resulting in unique outcomes in different geographical areas, whether in urban, urban/rural or rural contexts.

Figure 1.1 in Chapter 1 provides a framework for analysis of glocalisation processes. In tourism, the factors of transformation, hybridisation, accommodation and revitalisation traverse separate yet interconnected informal, hybrid informal/informal and formal economies. In this volume, the theorisation extends the understanding of glocal processes regarding *experience and behaviour of indviduals, place and identity and sustainable tourism.*

Various chapter contributions demonstrate the interaction between the global and the local, illustrating in the souvenir context how the local object and memory is transformed through acquisition and use into highly personal glocal forms. (Chapters 2–12) include related theories of self-image, occupational souvenirs, transformation, acculturation and marginalisation of authentic indigenous crafts, construction of place and cultural identity, marginalised and alternate economies at remote peripheries, transitions of informal to hybrid and entrepreneurial economies; all factors that link back into the research theory.

From the margins – people and places

Employing themes of people and places at the margins creates a context for the analysis of the meanings of souvenirs, not at the centre of the globalised world in a developed urban context, but at the edges of society, geographically and in terms of individuals and populations.

Peripheral locations are frequently distant from the urban centres of governance, have a small population and are often rural. These areas, with distinctive cultures preserved and protected by their distance from the centre and extreme physical locations, are impacted by globalisation forces (Katz, 2004). There has been considerable attention in the tourism literature to issues related to tourism development in peripheral locations and regions that include preservation of local cultures and the impacts of cultural change (Hohl & Tisdell, 1995). Development of tourism in these areas changes the relationship between people, place and tradition, as interactions between guests and hosts are embodied in the service encounter, in the case of souvenirs in the purchase transaction. Tourism brings consumers looking for authentic souvenirs reflecting place, and even for encounters with the makers of souvenirs. This demand creates an opportunity for local entrepreneurship and employment related to the production and sale of mementos to visitors. However, while stimulating the demand, in particular for art and craft production, the souvenir market can threaten authenticity in the items produced. While peripheral area tourism stimulates tourism demand, it includes the consideration of seasonality with related human resource management issues of labour, associated local economic development and cultural entrepreneurship (Commons & Page, 2001). However, there has also been acknowledgement of limited resources for the implementation of tourism policies and strategies in these areas (Albrecht, 2010) that may especially relate to nurturing local cultural production for tourism.

Groups on the margins of society also play a role in the realm of tourism souvenirs, for example for native and indigenous populations (Butler & Hinch, 2007). Involvement of small scale local producers (Swanson & Timothy, 2012) in the tourism trade of souvenirs has advantages and disadvantages, possibly creating employment and improving livelihoods, but at the same time, owing to the demands of global tourism, threatening the authenticity of indigenous arts and crafts and creating new levels of social interaction with visitors (Littrell *et al.*, 1993). Tourism, through souvenirs, can consequently act as an agent of cultural change in these situations. Local societies therefore performatively create alternate economies based on cultural processes and socially just practices (Gibson-Graham & Roelvink, 2011).

Cultural change

Tourism souvenirs are reflections of local cultural change in the context of globalisation. In this sense, souvenirs reflect the current trend to find meaning through tourism experience (Littrell *et al.*, 1993). The acquisition of souvenirs through travel is a cultural quest, centred on the fundamental principles of exchange between peoples, and an expression and experience of culture. In a changing glocal world, tourism souvenirs can represent the interplay between multiculturalism, retailing and even ethnic identity (Ahmad, 2003). Souvenir construction at the geographical and marginal peripheries of the globe, such as small island developing state locations, while influenced by global markets and forces may also support local craft traditions, livelihoods and economies (Hampton, 2005). The interplay of culture and change has led to tourism mobilising a multitude of forms of cultural expression (Robinson & Phipps, 2009), and in both intangible and tangible forms, cultural change is therefore embodied in the form of souvenirs, produced as commoditised objects or serving as evidence, acquired by visitors as markers of experience.

Research Gaps and Evidence

A number of academics have researched in the area of tourism and souvenirs, as noted in Chapter 1, from a variety of viewpoints that include perspectives on consumption patterns through retail shopping and tourist shopping villages; authenticity; and the role of, and impacts on, ethnic and tourist arts and crafts. Other academics addressed the role of souvenirs in society and the formation of identity; the purchase experience and behaviour; and their import to local economic and cultural development. Yet only several authors have identified the role of souvenirs in peripheral areas, while others have examined their relevance to community development.

Swanson and Timothy (2012) identified research gaps in the souvenir literature as including best practices, and internet opportunities, noting emerging areas of research as the salience of place attachment; ethnicity and nationality studies; retail analysis and merchant analysis. Wilkins (2011), stated a need for further research into purchase motivations and use of souvenirs, as opposed to the previous body of research in the literature focusing on types, uses and functions of souvenirs. The author observed that his findings provide opportunities for further research on the multiple roles of the souvenir as evidence, in particular as status symbols, consumption and an aid to memory.

Experience and behaviour

In a glocal context the contemporary tourism product is experiential, as tourists visiting local areas seek out life experiences and quality of life improvements through tourism (Carmichael, 2006). It is therefore useful to understand not only the tourism experience related to souvenirs, but also the behaviour towards souvenir sourcing, purchase and use. Tourists seek out souvenirs that are characteristic of the place that they are visiting (Gordon, 1986). This may include material objects with logos or pictures related to the place, or local foods that are representational of place. It is also known that tourists value the interaction with locals as they purchase souvenirs (Kim & Littrell, 2001). Contributions (Chapters 2–4) provide new insights into role of practical and occupational objects as souvenirs and the intangible consequences of souvenir acquisition as representative of experience.

Place and identity

Williams (1998) indicates that places, and images of places, are essential to the practice of tourism. There is sufficient evidence from the preceding chapters that souvenirs are inextricably connected to the images of places, either during the visit of the tourist, or after they have returned home. However, as the images produced as souvenirs for tourism purchase and consumption are subject to the forces of glocalised economies, traditional arts and crafts may be commodified or altered for the purpose of sale to tourists. Contributions (Chapters 5–7) consider new evidence regarding theorisations of place and identity in relation to souvenirs, especially in the case of indigenous groups, commodification and appropriated authenticity, as well as networks of supply.

Case studies

As Beeton (2005: 37) observes, case studies are extensively used in tourism research, having the advantage of being suitable for the more 'qualitative hypothetical deductive' and the 'holistic-inductive' paradigms of tourism research. Jennings (2001) notes case studies demonstrate a flexibility not obvious in other research modes.

The case study evidence has much to offer, not only for academics, but also as best practices that could be applied to other situations, addressing a research gap identified by Healy (1994). For example, there are instances when the production of souvenirs can strengthen local societies and economies, acting as a vehicle for the preservation of the local countering the pressures of globalisation, thus championing the preservation of local cultures and traditions. A number of international case studies (Chapters 8–12) provide evidence of best practices in sustainable tourism related to souvenirs, including the role of souvenirs in empowering local populations (Japan), the potential of online retail for stimulating post-visit consumption (Australia), the constraints and complexities peripheral communities face when entering the souvenir market (Tanzania and Thailand), transitioning of traditional villages from souvenir production to attraction (Vietnam) and the role of domestic tourism in stimulating souvenir production and consumption (Macau, PRC).

Discussion

In terms of theorising experience and behaviour, the chapter contributions expand the notion of a souvenir and extend the analysis of meanings attached to souvenirs. The notion of a souvenir, as a material object produced and then purchased as a memento, is extended to the idea of a souvenir as an object purchased for practical purposes, to which a souvenir meaning is attached and to occupational objects. Theorising culture, place and identity examined the socio-cultural processes in play in the production of souvenirs delving into the roles of producers and retailers. Authenticity is central to these discussions, as the souvenirs produced and then marketed in a glocal encounter are subject to forces that marginalise the original, authentic and real, as the objects produced are compromised by the pressures of production, as material culture such artefacts are even marginalised by the cultures that produce them. A critical turn on the examination of souvenir encounters, exposes the role of retailers in the portrayal of place and identity.

Souvenirs can be viewed as a tangible and often material aspect of the glocal encounter between the producer, the supplier, the retailer and the consumer. As such, souvenirs are evidence of the encounter, and their study has much to offer to add to our knowledge of, and insight into, glocalised tourism. Using a glocal lens has allowed for the preparation of concrete case studies that provide insight beyond the existing literature on tourism souvenirs into the struggles of the makers of souvenirs (for example Pacific migrants; rural Japanese, Thai, Tanzanians, Vietnamese and Laotians), as they deal with cultural and social contexts that often marginalise the authenticity and the viability of what they are producing for souvenir markets. This work has extended the boundaries of the research on souvenirs to include key themes of glocalisation, geographical peripheries, marginalised groups and societies, and alternative economies in terms of souvenir production and consumption. The book reflects upon current thinking and analysis on the theme of souvenirs, situating souvenirs as both tangible and intangible representations and evidence of tourism experience reflective of cultural change that are developed glocally, yet mirror global and local economic, political and societal forces.

The leveraging of real and imagined transnational diasporas (Croucher, 2012; Vertovec, 2001) to extend the 'effective national footprint' for trade, expertise and investment is a growing trend, vide New Zealand's KEA network (Gamlen, 2012) and Scotland's 'Homecoming', that, if incorporated in developmental thinking in glocal contexts, has the potential to develop sustainable income streams such as tourism (Coles & Timothy, 2004; Hall & Rath, 2007; Newland & Taylor, 2010; Nurse, 2011; Wickramasekara, 2009) for example through visiting friends and relatives (Cave & Hall, n.d.). Further, the work has explored the role of tourism in local economic development and cultural entrepreneurship, social and cultural preservation for communities and entrepreneurs located on islands, on urban and rural peripheries around the globe and for groups (self) located at the margins of society. It has pulled the research focus on souvenirs away from the dominant view of the souvenir as artefact and material culture, and the tourist as purchaser of a souvenir. It offers new perspectives of how the souvenir is experienced once the visitor has returned home, how the souvenir uniquely reflects place and identity and how the production of souvenirs by locals at the margins of the globe and society embodies the glocal, created locally but influenced by and situated within global forces, and has the potential to be foundational to new, diverse and perhaps more sustainable, economic forms.

Conclusion

The volume has provided a grounded framework for a new theorisation regarding souvenirs, extending the notion of the souvenir, the meanings associated with it and the activity of acquisition. It shifts the focus of research on tourism souvenirs from purchase and intent to meaning, evidence and use, not only in terms of the attainment of souvenirs to a societal and transnational contexts, but related to their production, consumption and authenticity. It acknowledges and addresses gaps in the previous research regarding souvenirs, especially those relating to place identity and best practices.

More research is needed on key areas introduced in this book in relation to souvenirs in terms of peripheral and marginal communities and groups, place identity and societal cultural change. Research on the impacts of transnationalism and diasporic creation and consumption would provide fertile directions for new analysis based on the theorisations introduced in this book. Geographically, the glocal context for souvenir production and consumption could be addressed in the context of other types of physical areas. Findings extending the 'notion' of the souvenir to practical items purchased during travel and to include occupational souvenirs of chefs could, in particular, be extended to other situations and occupational groups. Future research can thus continue to extend the boundaries of enquiry on tourism souvenirs and experience to the many glocal contexts in which souvenirs are produced, marketed and consumed.

References

Ahmad, J. (2003) Retailing in a multicultural world: The interplay of retailing, ethnic identity and consumption. *Journal of Retailing and Consumer Services* 10 (1), 1–11.

Albrecht, J.N. (2010) Challenges in tourism strategy implementation in peripheral destinations: The case of Stewart Island, New Zealand. *Tourism and Hospitality Planning & Development* 7 (2), 91–110.

Anderson, L. and Littrell, M.A. (1995) Souvenir purchase behaviour of women tourists. *Annals of Tourism Research* 22 (2), 328–348.

Beeton, S. (2005) The case study in tourism research: A multi-method case study approach. In B.W. Ritchie, P. Burns and C. Palmer (eds) *Tourism Research Methods: Integrating Theory With Practice*. Oxfordshire, UK: CABI Publishing.

Butler, R. and Hinch, T. (2007) *Tourism and Indigenous Peoples: Issues and Implications*. London: Routledge.

Carmichael, B.A. (2006) Linking quality tourism experiences, residents' quality of life, and quality experiences for tourists. In G. Jennings and N.P. Nickerson (eds) *Quality Tourism Experiences* (pp. 115–135). Boston, MA: Elsevier Butterworth-Heinemann.

Cave, J and Hall, C.M. (n.d.) Are today's migrants tomorrow's tourists? Sustaining tourism in the Pacific using diasporan networks. Paper submitted to *Journal of Travel Research*, Special Issue on the Future of Tourism in the Asia and Pacific region (under review).

Chang, T.C., Milne, S., Fallon, D. and Pohlmann, C. (1996) Urban heritage tourism: The global-local nexus. *Annals of Tourism Research* 23 (2), 284–305.

Coles, T. and Timothy, D.J. (eds) (2004) *Tourism, Diasporas and Space*. London: Routledge.

Commons, J. and Page, S. (2001) Managing seasonality in peripheral tourism regions: The case of Northland, New Zealand. In T. Baum and M. Landtorp (eds) *Seasonality in Tourism: An Introduction* (pp. 153–172). Oxford: Pergamon.

Croucher, S. (2012) Americans abroad: A global diaspora? *Journal of Transnational American Studies* 4 (2), 1–33.

De Sousa Santos, B. (2004) The world social forum: Toward a counter-hegemonic globalisation (Part I). In *World Social Forum: Challenging Empires*.

Gamlen, A. (2012) Creating and destroying diaspora strategies: New Zealand's emigration policies re-examined. *Transactions of the Institute of British Geographers* 36 (2), 238–253. DOI: 10.1111/j.1475–5661.2012.00522.x

Gibson- Graham, J.K. (2006) *The End of Capitalism (As We Knew It): A Feminist Critique of Political Economy* (1st edn). Minneapolis, US: University of Minnesota Press.

Gibson-Graham, J.K. (2008) Diverse economies: Performative practices for 'other worlds'. *Progress in Human Geography* 32, 613–632.

Gibson-Graham, J.K. (2010) Post-development possibilities for local and regional development. In A. Pike, A. Rodriguez-Pose and J. Tomaney (eds) *Handbook of Local and Regional Development*. London: Routledge.

Gibson-Graham, J.K. and Roelvink, G. (2011) The nitty gritty of creating alternative economies. *Social Alternatives* 30 (1), 29–33.

Gordon, B. (1986) The souvenir: Messenger of the extraordinary. *Journal of Popular Culture* 20, 135–146.

Hall, C.M. and Rath, J. (2007) Tourism, migration and place advantage in the global economy. In J. Rath (ed.) *Tourism, Ethnic Diversity and the City* (pp. 1–24). New York: Routledge.

Hampton, M.P. (2005) Heritage, local communities and economic development. *Annals of Tourism Research* 32 (3), 735–759.

Healy, R.G. (1994) Tourist merchandise as a means of generating local benefits from ecotourism. *Journal of Sustainable Tourism* 2 (3), 137–151.

Hohl, A.E. and Tisdell, C.A. (1995) Peripheral tourism: Development and management. *Annals of Tourism Research* 22 (3), 517–534.

Jennings, G. (2001) *Tourism Research*. John Wiley & Sons Australia Limited.

Kaplan, C. (1996) *Questions of Travel: Postmodern Discourses of Displacement*. Durham: Duke University Press.

Kim, S. and Littrell, M.A. (2001) Souvenir buying intentions for self versus others. *Annals of Tourism Research* 28 (3), 638–657.

Katz, C. (2004) *Growing up Global: Economic Restructuring and Children's Everyday Lives*. Minneapolis: University of Minnesota Press.

Littrell, M.A., Anderson, L.F. and Brown, P.J. (1993) What makes a craft souvenir authentic? *Annals of Tourism Research* 20 (1), 197–215.

Milne, S. and Ateljevic, I. (2001) Tourism, economic development and the global-local nexus: Theory embracing complexity. *Tourism Geographies* 3 (4), 369–393.

Newland, K. and Taylor, C. (2010) *Heritage Tourism and Nostalgia Trade: A Diaspora Niche in the Development Landscape*. Washington DC: Migration Policy Institute.

Nurse, K. (2011) Diasporic tourism and investment in Suriname. *Canadian Foreign Policy Journal* 17 (2), 142–154.

Pieterse, J. N. (1994) Globalisation as hybridisation. *International Sociology*, 9, 161–184.

Robinson, M. and Phipps, A. (2009) Are we there yet? *Journal of Tourism and Cultural Change* 7 (1), 1–4.

Rutten, R. (1990) *Artisans and Entrepreneurs in the Rural Philippines: Making a Living and Gaining Wealth in Two Commercialised Crafts*. Amsterdam: VU University Press.

Swanson, K.K. and Timothy, D.J. (2012) Souvenirs: Icons of meaning, commercialization and commoditization. *Tourism Management* 33, 489–499.

Timothy, D.J. (2011) *Cultural Heritage and Tourism: An Introduction*. Bristol: Channel View Publications.

Vertovec, S. (2010) *Cosmopolitanism Diasporas: Concepts, Identities, Intersections* (pp. 63-69). London: Zed Books.

Wickramasekara, P. (2009) Diasporas and Development: Perspectives on Definitions and Contributions. Perspectives on Labour Migration Working Paper No. 9. Geneva: International Labour Office.

Wilkins, H. (2011) Souvenirs – What and why we buy. *Journal of Travel Research* 50 (3), 239–247.

Williams, C.C. and Nadin, S. (2010) Entrepreneurship and the informal economy: An overview. *Journal of Developmental Entrepreneurship* 15 (4), 361–378.

Williams, S. (1998) *Tourism Geography*. London: Routledge.

Index

acculturation, 6, 15, 25, 77, 191
Adelaide, Australia, 16, 22, 136
Africa, 2, 23, 24, 32, 36, 151, 159
aide de memoire, 46
Akha, 150
alternate economies, 8, 192–193
alternative economies, 8, 21, 190, 196, 198
ambient variables, 135
American Indian nations, 65
American Southwest, 63
American West, 73
Americas, 55, 110
Anasazi, 66
Angkor World Heritage Site,
    Cambodia, 177
Anglo–European, 114
Aotearoa New Zealand, 98
Apache, 15, 63, 66
Appadurai, A., 4, 5–7, 18, 19, 84–86, 93, 96
Arab, 152
Arizona, USA, 64–65, 78, 80
artefact, 50, 52, 54, 57–58, 196
arts and crafts, 2, 3, 6, 20, 64–65, 73, 79,
    86, 126–128, 136, 161, 176, 183, 185,
    192–194
Arumeru District, Tanzania, 151, 153
Arusha National Park, Tanzania, 151
Arusha Town, Tanzania,151
Arvidsson, A.,12, 19
Asia, 2, 19–20, 24, 50, 55, 66, 73, 79, 84,
    88, 96, 150, 175–179, 186, 188, 197
Auckland, New Zealand,104
Australia, 16, 19–20, 22, 25, 42, 47, 50,
    54, 58, 104, 113, 133, 136–146, 160,
    195, 198

Australian, 22, 55, 57, 93, 104, 133,
    135, 146
authenticity, 1, 2, 15–19, 21, 24–25, 28,
    39, 41, 59, 64, 71–81, 91, 99–100,
    105, 112–116, 127–130, 158–159,
    164, 176, 182, 184, 192, 196

Bat Trang, Vietnam, 161–174
Bay of Plenty, New Zealand,16, 103–116
Belgian, 55
Bering Straits Land Bridge, 66
Boller, F., 12, 19
Brisbane, Australia, 16, 54, 136
British, 20, 24, 55, 115, 183, 198
Buddha, 15, 82–95
Buddhism, 88
Buddhist, 10, 83, 88, 92–93, 96

Camden Town, London, UK, 33
Canada, 22, 39, 78, 130, 133–134, 136
Canyon de Chelly, USA, 65
Captain Cook, 103
case studies, 150, 153, 160, 162, 165, 190,
    194–195
Cave, J., 1, 3, 8, 11–17, 19, 25, 77, 79,
    98–103, 114, 185, 187, 189, 196–197
celebrity chef, 49
chefs' uniform, 50, 53, 57
Chiang Mai, Thailand, 96–97, 150–151,
    153, 156
China, 17, 20, 24, 82, 110, 113, 121, 134,
    145, 167, 169, 177, 179, 180–181,
    183–184, 187–188
Chinese, 17, 49, 89, 151, 157, 172, 179,
    180–183, 185–187, 188

Choi Heong Yuen Bakery, Macau, PRC, 180, 187
Coastal Bay of Plenty, New Zealand, 16
Cohen, E., 2–3, 19–20, 102, 114–115, 120, 130, 147–150, 159, 163–164, 171, 174, 176, 186–187
colonialism, 84
colonisation, 15, 63, 76
Colorado, USA, 64
Colorado Plateau, USA, 64
Colorado River, USA, 65
commodification, 2–6, 16, 23, 75, 79, 99–100, 105, 113, 120, 133, 176, 194
constructed authentic, 15
constructive authenticity, 73, 78
core-periphery relationship, 148
Croatia, 33
cruise ship market, 104
cultural change, 1, 25, 190, 192–193, 196–197
cultural entrepreneurship, 1, 192, 196
cultural forms, 6
cultural identity, 5, 11–12, 63, 77, 98, 107, 181–182, 186, 192
cultural receptivity, 6, 14
Cultural Revolution, 184
cultural tourism, 19, 74, 177, 179, 186–187
curio business, 73

de Beauvoir, S., 10
decay of acquisitiveness, 14
design variables, 135
diasporas, 22, 23, 196
du Cros, H., 17, 23, 176–177, 179, 186–187

East Asia, 5, 25, 84, 96, 182, 187
eco-museums, 119
economic development, 5, 18, 23, 63, 64, 80, 148, 152, 158, 160, 174, 192, 196, 198
Ecuadorian, 55
El Tovar, USA, 73
English, 22, 101, 116, 153, 167, 181
enterprise development, 3, 11
ethnic arts, 121, 131
ethnography of commodities, 15, 83–85
Eurocentric, 82

Europe, 2, 53, 55, 104, 185
European, 24, 50, 60, 66, 84, 93, 101, 103, 114, 146
evidence of experience, 189, 191
export development, 6

farmers' markets, 16, 119, 122–123, 126, 129
first-world, 63, 64
food souvenir shops, 183
Four Corners Region, USA, 64
fourth world, 63
Fred Harvey Company, 73, 79–81
French, 53, 55, 84, 88–89, 96, 101
Friends of Vietnam, 174

Ganado Trading Post, 68
German, 55, 136, 152, 162, 169, 174
German Technical Cooperation, 162, 169, 174
Germany, 59, 127
Gibson-Graham, J.K., 4, 8, 198
global manufacturers, 3
globalisation, 1, 3–4, 6, 28, 64, 149, 157, 159, 173, 190, 192–193, 195
globalised localisms, 190
global–local relationship. See glocal
glocal, 173, 189, 190–91, 193–197
glocalisation, 1, 4–6, 13–14, 55, 99, 100, 190–191, 196
Goffman, E., 52, 58–59
Gordon, B., 2, 21, 28, 30, 39–40, 47, 51, 68, 79, 101–102, 110–111, 115, 127, 130, 171, 174, 189, 194, 198
Gordon Ramsay, 51
Grand Canyon, USA, 65, 73
Green Tourism, 16, 119, 121–123, 127–131
Griffith University, Australia, 42
grobal, 6, 14
Guangdong, PRC, 179, 182

Hahndorf, Australia, 133, 136–137, 145
Hakataya, 66
Hampton, M., 149, 160, 164, 174, 193, 198
handicraft villages, 17, 161–162, 164–165, 167, 170–173, 175

handicrafts, 6, 8, 22, 102, 116, 120–121, 131, 148, 149, 162–163, 165, 171, 182, 184–185
Hanoi, Vietnam, 165–167, 171, 173–175
Hashimoto, A., 3, 16, 21, 68, 79, 102, 105, 115, 119, 120–121, 127, 129–130
Havasupai, 66
heritage tourism precinct, 177
heritage-tourism nexus, 84
hill tribe tourism, 150
Hitchcock, M., 2, 21, 28, 39, 83, 95–96, 120, 131, 176, 186–187
Hmong, 150
Hohokam, 66
Hoi An, Vietnam, 165, 168–174
Hollywood image, 78
Hong Kong, PRC, 6, 19, 178–179, 181–183, 186–187
Hopi, 67
Huay Pu Keng, Thailand, 150–151, 155–156, 158
Hungarian, 55

imported souvenirs, 150, 157–158, 169
Indian Arts and Crafts Act of 1990, USA, 75
Indian City, USA, 76
Indian Ocean, 152
indigenous cultures, 63
indigenous peoples, 64
informal economy, 8–9, 13, 16, 25, 99, 160, 199
internet connective technologies, 191
interpretation of place, 145
Irish, 55, 79
Iron Chef, Fuji Television, Japan, 49
island contexts, 3
islands, 2, 103, 131, 196
Israel, 31, 34

Jamie Oliver, 51
Japan, 8, 16, 23, 49, 96, 119–123, 127, 129–131, 162–163, 171–172, 174, 180, 195
Japan International Cooperation Agency, 162, 163, 171–172, 174
Japanese, 49, 91, 119, 123, 126–127, 129, 180, 183, 195
Japanese Ministry of Agriculture, Forestry and Fisheries, 129

Kakaji Village, Japan, 123
Karen, 150, 151
Katikati, New Zealand,106
Kim Bong, Vietnam, 165, 168–174
Korean, 49
Kunisaki Peninsula, Japan, 119, 121–122, 126, 129
Kyushu Island, Japan, 119

Lahu, 150, 155
Lao People's Democratic Republic, 83
Lao textiles, 82, 89, 96
Laos, 82–83, 88–89, 91, 96–97
Latin America, 22, 121
lifestyle migrants, 6
Lisbon, Portugal, 185
Lisu, 150, 155
Little Colorado River, USA, 66
Littrell, M.A., 2–3, 18, 22–23, 25, 28, 30, 37–41, 46–48, 67, 71, 79, 99, 100, 102, 114–116, 158, 160, 164, 174–175, 189, 193–194, 197–198
local artisans, 145
local products, 73, 102, 105, 111–113, 126, 136, 139, 142–143, 145, 169
local–global, 4, 15, 19, 21, 49, 83, 105, 113, 115,
Lonely Planet, 157
Louis Akin, 73
Luang Prabang, 15, 82–96
Luang Prabang Buddha sculptures, 93
Lury, C., 2, 22, 102–103, 110–111, 115

Macanese, 181, 185
Macau, PRC, 17, 176–188, 195
MacCannell, D., 40, 71, 76, 79, 82, 96
Mae Aw, Thailand, 151
Mae Hong Son, Thailand, 151, 153, 156
Magokoro Farmer's Market, Japan, 123
Maji Ya Chai, Tanzania 151
Maori, 103, 108
marginal communities, 197
marginalisation, 11, 14, 22, 24, 63–64, 73, 76–78, 191
marginality, 9–11, 14–15, 58
markers (souvenirs), 54, 57, 102–103, 111, 127, 193
Matama, Japan, 123

material culture, 1, 15, 17, 18, 22, 77, 82–84, 86, 95, 101, 105, 112, 176–177, 188–189, 195–196
Mayan craftswomen, 120
McDonalds Restaurants, 5, 55
McKercher, B., 3, 11, 23, 59, 176, 188
Melbourne, 22, 33
memento of travel, 83
mementos, 14, 17, 30, 32–33, 36, 40, 103, 176, 183, 192
Mesa Verde, USA, 68
Mexico, 73
micro-finance, 156
Micronesia, 121, 131
Middle East, 55
Mien, 150
migration, 5, 9, 11, 21, 25, 115, 134, 151, 163, 198
Ministry of Agriculture and Rural Development, Vietnam, 162
Mitchell, K., 7, 23, 134, 145–146
mobilities, 7, 83, 95, 98, 103, 105, 108, 114
mobility, 1, 5, 15, 51, 58, 65, 85, 103, 105, 107, 113
modernity, 17, 19, 73, 95–96, 177, 186, 190
Mogollon, 66
Montville, Australia, 133, 136–137
Morgan, N., 23, 28, 30, 39, 101–102, 116, 184, 188
Mount Kilimanjaro International Airport, Tanzania, 151
Mount Maunganui, New Zealand, 104, 106, 109
Mount Meru, Tanzania, 151
Murphy, L., 2, 6, 11, 16, 23, 51, 59, 132–137, 146
Murphy, P., 11, 23
Muslim, 152

Native American, 63–64, 67–69, 72–80
Native American tribes, 63
Navajo, 15, 63, 65–70, 73, 76–77, 79–80
New England, USA, 134
New Mexico, USA, 64–66, 79–81
New York, USA, 33
New Zealand, 19, 23, 39, 55, 78, 98, 103–105, 110–116, 133, 136, 159, 187, 196–198

Ngongongare, Tanzania, 151
Nguruma, Tanzania, 151
Niagara Region, Canada, 127
North Island, New Zealand,103–104
notion of a souvenir, 195
Nud*ist, software program, 42

Oaxaca, Mexico, 120
objective authenticity, 71
objectively authentic, 15, 64, 72, 74, 78
objects of travel, 102
occupational artefacts, 15, 49, 51–58
occupational jurisdiction, 57
Oita, Japan, 121
online retail, 145
Oriental cuisine, 50
Orientalism, 11, 96
Orthodox Church, 52
outlet mall, 134, 141

Pacific Islands, 2, 104
Pangani District, Tanzania, 152
Pangani, Tanzania, 153
Patandi, Tanzania, 151
peripheral communities, 17, 147–148, 195
peripheral economies, 150, 159
peripheries, 1, 2, 9, 11–14, 18, 22, 149, 162, 164, 171–173, 190, 192–193, 196
periphery, 163
Philippines, 8, 23, 163, 175, 198
physical artefacts, 101
pictorial images, 102, 111
piece of the rock souvenirs, 102
Pima, 66
Pong Ngean, Thailand, 151
Portuguese, 66, 179, 181, 183–187
postmodernist, 177
post-structural approach, 105
poststructural critical turn, 16
Prefecture of Oita, Japan, 119
Pritchard, A., 23, 28, 30, 39, 101–102, 116, 184, 188
product placement, 6
Pueblo, 15, 63, 66, 68–69, 71, 73–74, 76, 79–80
Pueblo culture, 65
purchase of souvenirs, 41, 134
purchase opportunity, 144

purchase transaction, 192
purchasing behaviour, 3

Quang Nam Small and Medium Enterprise and Cooperative, 170

remembrance, 83, 94, 116
retail shopping, 2, 146, 193
Rio Grande River, USA, 65, 66
Ritzer, G., 4, 6–7, 24
Robertson, R., 4–6, 14, 21, 99, 114–115
role of women, 120
Rotorua, New Zealand, 104

Saadani National Park, Tanzania, 152
Saange, Tanzania, 152
Said, E., 84, 96
San Juan Region, New Mexico, 66
Santa Fe Railroad, 73, 81
shopping, 6, 20, 23, 25, 28, 40–41, 46–47,
    102, 104, 115–116, 132, 134–135,
    137, 139–141, 143–146, 148, 154,
    164, 167, 174–175, 179, 184–185, 187
shopping and souvenirs, 6, 148
shopping expenditure, 40, 46, 135
shopping precinct, 141
Singaporean, 55
social capital, 85, 158
social identity theory, 51
social variables, 135
social worlds theory, 51
socio-spatial characteristics, 6, 14
South African, 55
Southwest, USA, 63–69, 71–81
souvenir attributes, 42
souvenir consumption, 15, 80
souvenir craftsmen, 149
souvenir creation, 9, 189, 191
souvenir industry, 9, 128, 152
souvenir market, 17, 73, 148, 156,
    192, 195
souvenir purchase behaviour, 15, 40
souvenir purchase motivations, 46
souvenir purchases, 3, 47
souvenir shop, 7, 107, 126
souvenir studies, 190
souvenirs and tourism, 3
SpaLand Hotel and Resort, Japan, 123
SpaLand Seniors' Workshop, 126

St Lazarus Conservation Area, Macau,
    PRC, 184, 186
St Paul's Ruins, Macau, PRC, 177,
    180–181, 183–184, 187
Statistics New Zealand, 104
Sun Western, 123
Swahili, 152
Swanson, K.K., 2–3, 6, 8–9, 15, 24, 30,
    39, 63–64, 67–68, 71, 74, 77, 80, 98,
    101–102, 116, 120, 131, 133, 146,
    163–164, 175, 192–193, 198
Swyngedouw, E., 4
symbolic shorthand souvenirs, 102

Tanga Region, Tanzania, 152
Tanga City, Tanzania, 152
Tanzania, 16, 32, 148, 150–157, 160, 195
Taos, USA. 68, 73
Tashibunosho Village, Japan, 123
Tashibunosho Women's Group, 126
Tasmania, 178, 187
Tauranga, New Zealand, 104
Te Moana a Toi Bay of Plenty, New
    Zealand, 103–107, 110
Te Puke, New Zealand, 106
Teague, K., 2, 21, 39, 83, 95–96, 187
Telfer, D.J., 3, 16, 21, 68, 79, 102, 105,
    115, 119, 120–121, 127, 129, 130
Thailand, 16, 19, 20, 22, 25, 92–93, 148,
    150–156–157, 159–160, 163–164,
    171, 174, 177, 187, 195
thanatourism, 101
The Culinary Institute of America, 52, 60
Thomas Moran, 73
Timothy, D.J., 2–3, 6, 8–9, 24–25, 28, 39,
    67–68, 71, 74, 80, 98, 101–102, 116,
    120, 130–131, 133, 146, 163–164,
    175, 178, 187–189, 192–193, 196–198
Tokyo, Japan, 121
Tom Lod, Thailand, 151, 155
Tonga, 8, 18, 121, 130
touchstones of memory, 101
Tourism Department of Quang, Vietnam,
    170
tourism imaginaries, 6
tourism industry, 6, 11, 46, 65, 98, 103,
    106, 121, 152–153, 156, 158, 169,
    171, 190–191
tourist keepsakes, 83

tourist mementos, 3, 101
tourist-objects, 102–103, 109–111
tourist shopping, 2, 16, 23, 132, 134–136, 143, 146, 164, 173, 193
tourist shopping villages, 2, 16, 132, 146, 193
tourist souvenirs, 65, 67–68, 70, 73–75, 120
touristic trinket, 49
tourists arts, 2
transnationalism, 191
travel histories, 31
travel motives, 31
traveller, 12, 49–50, 76, 83–84, 89, 95, 103–104, 161, 167, 171, 173
traveller-objects, 102–103, 108–111
tripper-object, 102–103, 109–111

UK, 21–22, 24, 39, 55, 80, 115, 197
UNESCO, 168, 170, 178, 184
United Kingdom, 2, 104, 107, 136
United States, 2, 20, 63–64, 73, 79, 133, 136, 159
Urry, J. 22, 65, 80, 85, 96–97, 103, 115–116
Usa River Widows Group, Tanzania, 151
Ushongo, Tanzania, 152
Uswahilini, Tanzania 151
Utah, USA, 64

Van Phuc, Vietnam, 166

Vietnam Administration of Tourism, 161, 167, 170, 172–173, 175
Vietnam, 17, 161–163, 165, 167–171, 173–175, 195
Vietnam Handicraft Research and Promotion Centre, 162
Visitor centres, 3

Waikato, New Zealand, 104, 114, 187
Waitomo Caves, New Zealand, 104
Wat Visounnarath, Luang Prabang, Laos, 86
Wat Xien Thong, Laos, 88–89, 91–92
Western, 15, 23, 50, 65–66, 82, 116, 123–124, 177, 180–182, 185–186
Western Canada, 66
Wilkins, H. 3, 6, 14, 25, 40, 41, 44, 48–49, 58, 60, 132–133, 146, 164, 171, 175, 189, 193, 199
Wood, R.C., 49, 51–52, 55–56, 60, 108, 184
World Heritage List, 84
World Tourism Organization, 147, 184, 188

Yavapai, 66
Yin, R.K., 153, 160, 164–165, 175

Zanzibar, Tanzania, 152
Zuni, 67